DYNAMICS OF AGRICULTURAL DEVELOPMENT

VOLUME II : TECHNOLOGICAL CHANGES AND SUSTAINABLE DEVELOPMENT

Vol. I : Land Reforms, Growth and Equity
Vol. II : Technological Changes and Sustainable Development
Vol. III : Policy Planning and Liberalisation

About the Authors

Dr. K.S. Dhindsa is a Professor of Economics at the Punjab School of Economics, Guru Nanak Dev University, Amritsar (India). Dr. Dhindsa has a brilliant career, who after graduating from Punjab University, completed the post-graduate diploma in Economic Development from Glasgow University. Thereafter, he obtained his Master Degree from Leeds University (England). Then he went to London University where he successfully completed his Ph.D. degree. He has published a book titled "India's Export Performance—Some Policy Implications". He has published a fairly large number of research papers in National and International journals. For his post-doctorate he went to the School of Oriental and African Studies, University of London, during 1989-91. He was the Editor of *Indian Journal of Quantitative Economics*—A Journal of International Development, during 1994-96. He was the Dean of Faculty of Economics and Business during 1992-93 and was Chairman of Punjab School of Economics, during 1994-96.

Anju Sharma is a Lecturer in Economics in the Post-Graduate Department of Economics, Khalsa College, Amritsar (Punjab). She obtained her M.Phil and Ph.D. degree from Punjab School of Economics, Guru Nanak Dev University, Amritsar. She has carried out a variety of research in the field of Economics. She is the author of a book entitled. *Dynamics of Cropping Pattern : A Regional Study of Punjab.* •
A dozen of her research papers have already been published in reputed Journals.

DYNAMICS OF AGRICULTURAL DEVELOPMENT

VOLUME-II

TECHNOLOGICAL CHANGES AND SUSTAINABLE DEVELOPMENT

K.S. DHINDSA
ANJU SHARMA

CONCEPT PUBLISHING COMPANY, NEW DELHI-110059

ISBN : 81-7022-836-0 (Vol. II)
ISBN : 81-7022-834-4 (Set)

First Published 2001

Published and Printed by
Ashok Kumar Mittal
Concept Publishing Company
A/15-16, Commercial Block, Mohan Garden
New Delhi-110059 (INDIA)
Phone : 5648039, 5649024
Fax : 091-11-5648053
Email : publishing@conceptpub.com

Foreword

Development of the agriculture sector in India has traversed a period of five decades since the country achieved Independence in 1947. During this period the very response system of governance has undergone a discernable change from governing an important colony to managing of an independent society that had an entity as a nation state, which was riddled with some formidable problems on economic, social, culture and political fronts. It was under this changed response scenario that agriculture sector assumed prime importance in the development process of the Indian economy so that (1) food and fiber needs of the society were met on a sustained basis, and (2) more than 80 per cent of the population living in rural areas who were wholly or partially dependent for their wherewithal on agriculture sector were provided with gainful employment opportunities. It is of course a different matter, the deriving inspiration from socialistic system of the then USSR, industrial sector was encouraged as a future in the development process and with the public sector achieving commanding heights, distributional social justice was aimed at for the populace on economic front. Yet, this approach and overt emphasis on industrial sector, specially the obssession with big industries, was not supposed to dilute the prominency of the agriculture sector; may be out of sheer compulsions of feeding the large and fast growing population of the country.

Thus, on one hand serious effort was made and are as yet afoot to tone up the agricultural sector for higher growth through strengthening the agricultural research system aimed at developing appropriate production technology, farm level extension of the new technology and agricultural education as well as institutional and policy support and on the other hand the agriculture sector was systematically squeezed to yield marketed surpluses, specially of foodgrain to feed the urban population through public distribution system. Thus, agriculture sector acted as a lead sector that yielded capital surpluses and wage good following a classic model of economic growth and development, though sluggish in pace, as postulated for agriculture sector dominated developing societies, this approach manifested through adoption of the process of investment planning and system of five year development plans.

The priority accorded to the agriculture sector was visible in the First Five Year Plan of 1951-52 through 1955-56. The explicit-emphasis, however, got diluted in the Second plan. Even the Third Five Year Plan did not absorb the reality of the situation and agriculture sector did not get adequate allocations. This happened because of good production response of the agriculture sector in the First Five Year Plan. By the time the mistake was realized it was too late and the country had virtually become a 'basket case' by the mid-sixties.

This impelled the policy makers to accord priority to the agriculture sector on technological improvement through expanding and strengthening the national agricultural research system, including the central and state level agricultural universities and research institutes. The research effort was complemented with reorganisation and toning up of extension network and also with reorganization of agricultural education. In order to boost sustained growth in agricultural production, supportive institutional and policy framework was put in position through a system of minimum support-prices and procurement of agricultural produce by the government. The creation of Food Corporation of India (FCI) and Public Distribution System (PDS) was a land mark in the history of Independent India.

The basic purpose of the system was three fold :

(i) Provision of price certainty to the agricultural producer at reasonable (remunerative) level to save him from vagaries of buyer's market in which stands helpless spectator in the post harvest period, when the buyers so often pooled to bring down the market prices to the disadvantage of the producer seller. With the introduction of the system of support/procure-ment prices, which by and by covered 22 commodities, the government stand obligated to lift whatever quantities the farmers offer for sale to the government at the announced prices. For this purpose Food Corporation of India acted and any loss was to be covered by the Central Government and profit was to be turned over to the government. The FCI acted direct as well as through the state agencies as their agents.

(ii) Supply of foodgrains, sugar, oil and raw material to the urban population at a reasonable prices through the Public distribution system fed through the procurement of foodgrains, specially wheat, rice and sugar appropriated through sugar levies on sugar mills. Raw materials like jute and cotton were procured from the market at support prices and also quite often at market prices through jute corporation of India (JCI) and cotton corporation of India (CCI) respectively for consumption of the public enterprises.

(iii) Creation of buffer stocks for food security because Indian sensetivities were deeply hurt in mid-sixties when it had become difficult for the government to buy foodgrain from the international market. In order to balance these three, often time conflicting, objectives, Agricultural Prices Commission (APC) was created, which was later renamed as Commission for Agricultural Costs and Prices (CACP), with expanded scope.

Initially the system worked well. In fact during first two years the FCI registered net profit. Yet, soon the ruling party's approach to appease the urban consumers resulted in subsidized issue prices at the PDS and foodgrain as well other commodities were supplied to all the consumers holding ration cards without targeting the distribution for the poor. As a consequence the budget provisions for food subsidies started escalating. Since the government opted not to increase issue prices for the consumers out of political reasons the brunt was borne by the farm sector. Government hesitated to increase support/procurement prices and turned the terms of trade against the farm sector. With a view to maximizing procurement of foodgrain at low prices (support/procurement prices) indirect approach was adopted which eliminated the private trader from the grain-markets. Drastically reducing the limit on food credit (selective credit), putting restrictions on holding of stocks, denying railways wagons to the traders and putting restriction on inter-state or inter-zone movement of grains the market prices were artificially brought down to the level of support/procurement prices. This process continued throughout during the past three decades with exception for only a few years intermitantly when farmers reaped bumper crops, stores/godowns of the government were full and consumer demand and liftings from PDS were low. These years too the farmers suffered on account of back of market clearance for their produce. Thus on one side the production was encouraged as a policy and on the other produce was procured for consumers on low prices to keep the urban voters happy. Thus a great deal of politics, rather cheap politics got mixed with the economics of the policy, which did not put the agriculture sector on an optimal growth path, put a drain on the budgetary resources of the government and created unwanted and unwarranted expectations in the minds of the urban consumers, which promoted pressure on the political system to keep consumption subsidized without any targeting for the poor.

As a consequence of these policy distortions, the agriculture sector remains net taxed or under negative protection according to some scholars. Although this view is widely accepted, yet there are other economists who believe that agriculture sector be brought under the income tax net. There is some logic in this argument too, because there are quite a large number of big farmer who earn huge incomes from farming, yet the sector as a whole is not capable of yielding any income tax, specially when collection costs are kept in view.

There are several other problems the agriculture sector faced as it progressed through last five decades and bore the impact of policy prescriptions at various stages of development. At this stage the farm sector is on crossroads, where production is not picking up as fast as is the need of the nation, weather as yet remains a very important factor and minor deficiencies in rain put the growth of the sector of the track, research and technology development has gone sluggish, policy has gone into a *rut*, farmers are coming under economic squeeze, consumers are becoming more demanding, population growth and improvements in money incomes are increasing the demand for food and fibers, government exchequer are coming under financial strain and above all due to progressive liberaliztion and globalisation of the agriculture market, the agriculture sector is going to face global competition. These are a few of the challenges that are staring at the face of the policy makers and market players in this sector.

It is against this background that the subject matter of this book assumes topical importance. The coverage of the subjects and issues like institution changes in historical perspective, reflections on input out put relationships and their impact on agricultural growth; impact of price support and market intervention, relationships between technology and growth as well as equity, implications and possibilities of agricultural taxation and changing terms of trade, impact of agricultural growth on labour absorption, specially with the adoption of improved technology, probable impact of liberalization and globalization on the agriculture sector as well as general review of macro and micro level agricultural policies followed through this period have a great relevance today if this sector is to be put on an optimum growth path in order to meet the fast growing demand for food and fibre in the national market.

This book is a good job done in this respect; which should prove to be a good material for students and researchers engaged in the field of agricultural economics and specially for those specializing in agricultural development policy.

S.S. JOHL
Prof. of Eminence in Economics (Retd.) and
former Vice Chancellor
Punjabi University, Patiala

PREFACE

There exists at present a considerable body of literature on Indian agriculture in different research publications. Some of this material is accessible in a few edited volumes, but none of them covers all the major, constantly swelling aspects of this subject. So the necessity for editing a comprehensive book on Indian agricultural development was being felt since long. The present book "Dimensions of Agricultural Development" is a humble attempt towards this end. In this book an attempt has been made to include good quality articles of renowned economists published in various reputed Journals. These studies, although related to India's agricultural development can be read with keen interest by various economists and policy-makers who are interested in agricultural Development of other third world countries.

While selecting various articles for this volume, syllabi of agricultural economics for post-graduate classes, prepared recently by various universities of India was also kept in view. This book will be of immense use to the students, teachers, researchers and policy makers in India as well as abroad.

In order to make a detailed study of various dimensions of agricultural development in India, thirty three articles have been included in three different volumes. Articles included in these volumes cover both micro and macro aspects pertaining to Indian agriculture, examine the issues and problems empirically and also provide careful presentation of most of the relevant facts of India's agricultural development.

The first volume of the book deals with the issues relating to land reforms, growth and equity. The second volume comprises of various articles relating to technological development in agriculture and sustainability of agricultural development. Third volume deals with the policy and planning concerning agricultural development and the impact of recently concluded GATT round on Indian agriculture.

The main introduction of the book has been given in the first volume. Most of the articles included in these volumes are reprinted articles and some of them are specially invited papers. Details of the original source of each articles has been given in the notes at the end of the chapter.

In an edited book, the co-operation of all those people is very much required whose work we wish to include. We must express our thanks to the contributors for giving us the permission to reproduce their research papers in this book. We are also grateful to the publishers of the Journals who have permitted us to reproduce these papers from their publications.

Before finalising the contents of the book, we had discussions with professors Yoginder K. Alagh, G.S. Bhalla, G.K. Chadha, Krishna Bhardwaj, S.S. Johl, K.L. Krishna, V .M. Rao, Suresh S. Tendulkar and Autar Singh Dhesi. We gratefully acknowledge their valuable suggestions. Thanks are also due to the staff members of Rattan Tata Library, Delhi, School of Economics, Institute of Economic Growth Library (Delhi), National Council of Applied Economic Research (Delhi), Jawaharlal Nehru University Library (Delhi), and Punjab Agricultural University Library (Ludhiana) for providing us facilities to use the relevant research material.

Last but not the least, we are profoundly grateful to the publisher M/s. Concept Publishing Company, of their whole-hearted co-operation in bringing out this book in record time.

K.S. DHINDSA
ANJU SHARMA

CONTENTS

LIST OF CONTRIBUTORS

Bina Aggarwal
Professor, Institute of Economic Growth, University Enclave, Delhi

Derek Byerlee
c/o Department of Economics and Sociology, Punjab Agricultural University, Ludhiana

G.K. Chadha
Professor, Centre of the Study of Regional Development, Jawaharlal Nehru University, New Delhi

Kanchan Chopra
Professor, Institute of Economic Growth, University Enclave, Delhi

K.S. Dhindsa
Professor, Punjab School of Economics, Guru Nanak Dev University, Amritsar

Romesh Diwan
Professor, Department of Economics, Rensselaer Polytechnic Institute, Tory, New York, USA

Robert E. Evenson
Professor, Economic Growth Centre, Department of Economics, Yale University, New Haven, USA

M.V. Srinivasa Gowda
Professor, Department of Economics, University of Agricultural Sciences, Hebbal Campus, Bangalore.

C.H. Hanumantha Rao
Professor, Centre for Economics and Social Studies, Hyderabad

S.S. Johl
Professor of Eminence (Rtd.), Department of Economics and Sociology, Punjab Agricultural University, Ludhiana.

Renu Kallianpur
c/o Department of Economics, Rensselaer Polytechnic Institute, Tory, New York, USA

Anju Sharma
Lecturer, Department of Economics, Khalsa College, Amritsar

R.K. Sharma
Reader, Centre for the Study of Regional Development, Jawaharlal Nehru University, New Delhi

D.S. Sidhu
Professor (Retd.), Department of Economics and Sociology, Punjab Agricultural University, Ludhiana

S. Suryaprakash
Professor, Department of Agricultural Economics, College of Agriculture, Mandya

M.S. Swaminathan
Chairman, M.S. Swaminathan Research Foundation, Institutional Area, Madras

J.V. Venkataram
Professor of Agricultural Economics, Alamaya University of Agriculture, Alamaya, Ethiopia

LIST OF TABLES

INTRODUCTION

K.S. DHINDSA & ANJU SHARMA

Agricultural technology refers to knowledge used in enhancing agricultural productivity. It points to the combination of inputs used and the improvements that occur in them from time to time with a view to enhancing productivity at the same unit cost or maintaining the current levels of productivity at a reduced cost. The selection of a suitable technology is very much important for an efficient production system. With the passage of time the nature of agricultural technology has changed gradually from its traditional stage to modern stage. The new farm technology which is currently being used in various advanced countries and also in major areas of the under developed countries is mainly the Seed, Irrigation, Fertiliser, Pesticide technology. New technology has played very important role. It benefits the farmers in terms of increased agricultural production and the consumers in terms of reduced prices for a better quality of agricultural output.

Keeping in mind the nature of agricultural operations, it is observed that modern technology enters agriculture sector through large-size holdings due to the fact that these holdings have the capacity to afford the use of mechanical power which is invariably an important component of new farm technology. Due to the high investment potential capacity, and other natural advantages, big farmers become the main achievours of gains terms of increased income. Small size farmers also follow with some lags but their income gains could not match, with the large farmers. This is one of the main reasons why the new farm technology has widened the income inequalities. In recent years, new farm technology has played an important role in enhancing global food production and to cater to the rising food requirements.

In modern times the package of inputs used in agricultural operations has changed a lot. In this package of inputs, human and animal labour is increasingly being replaced by the mechanical power, organic manures being substituted by chemical fertilisers and the traditional seeds are replaced by new high yielding varieties of seeds. Besides these components, pest control and scientific farm management are also concerned with the new farm technology. No doubt the success of new technology depends very much on the availability of adequate and assured irrigation and thus obstructs its, applicability in those areas where irrigation facilities are non-existent or not sufficient. But the development of artificial irrigation depends greatly upon the geographical and topographic conditions of a region. Realising the impossibility of providing assured irrigation facilities in all cultivable areas, the scientists have started concentrating on research in dry-farming methods. Important break through in this lime has still to be achieved.

Green revolution in India is the combined result of various measures taken by the government. The changes in cropping pattern witnessed during the post-green revolution period have shown some important effects. The shift of cropping pattern in favour of highly profitable crops like wheat and rice shows the rational behaviour of the farmers. The fast increase in assured irrigation facilities providing basic and necessary conditions for the cultivation of these crops supported by fast increase in their yield as a result of rapid adoption of high yielding varieties of seeds and use of adequate doses of fertilisers and pesticides have led to fast increase in area under cereals and other commercial crops. During Sixth and Seventh Five Year Plan, plant protection measures were strengthened to reduce crop losses especially in case of cotton, pulses and oilseeds.

Infact Green Revolution in India has been successful because of a strong infrastructural base built up by the government during the plans. Important constituents of infrastructure include transport and communication, regulated markets, storage and warehousing, agriculture education and training and agricultural research and extension. Fixation of minimum support prices, procurement prices and effective procurement operations carried out by the government has provided ready and assured markets to the farmers for the sale of various crops. No doubt machinery displaces labour and therefore leads to unemployment. But the new technology necessitates the use of machinery. Bumper harvest of wheat necessitates the use of threshers and the combined harvestors. The harvestors become necessary because these help in preparing the land for the next crop in time. For timely sowing, the tractors are also very much essential.

Although modern technology is very much helpful in India in achieving the target of self sufficiency in the production of foodgrains but the impact of this technology on the environmental situation is very much complex and manifold. The rice-wheat cropping system presently dominating in most of the States in India does not seem to be sustainable in the long run because the repetition of this crop rotation has led to considerable reduction in ground water balance, deterioration of soil health, multiplication of pests, diseases and weeds excessive use of electricity, fertilisers and pesticides and deterioration in the overall agro-economic system of the country. An increased use of chemical fertilisers has been hardening the upper crust of the soil which in turn destroys the moisture retention quality of the soil. Thus due to it the high lands do not retain moisture which is very much useful for living organism such as earth worms and other useful bacterias which inturn are highly conducive for the growth of crops.

The substantial increase in area under rice and intensive exploitation of groundwater resources through installation of tubewells have resulted in decline in water table in some States of India. This over-excitement of adopting new technology has made the farmers indifferent towards the future consequences and hence excessive use of modern technology. In some places, even the problem of salinity and waterlogging are repeated to be causing concern. In a country like India, where land is already very scarce, wasting some areas due to waterlogging can not be afforded very easily. Similar problems have also arisen with respect to those activities where forest resources are being used, excessive use of forest resources without giving proper attention towards their regeneration is also causing a great concern.

The above discussion clearly shows how agriculture and environment come into contradiction with each other. This contradiction has already led to different forms of environmental degradation like deforestation, air and water pollution besides other problems relating to soil health.

Keeping in mind the above mentioned adverse affects of new farm technology on environment, the need for protecting soil health and environmental degradation is strongly being felt in both developing and developed countries of the world. In Industrialised countries, the major concern is naturally the maintenance of the *status quo* in yield since this will be more than sufficient to meet their domestic and export requirements. But in developing countries like China and India, the challenge is how to produce more and more from diminishing land and water resources. This emphasis the need to consider methods of achieving continuous advances in agricultural productivity in an ecologically sustainable manner.

For achieving sustainable growth in productivity together with ecological sustainability, measures like effective management of available surface and ground water resources need to be taken up in an appropriate manner. Excessive use of irrigation water should be checked through volumetric pricing of water, which in case of electric tubewells means levying tariff on energy (electricity) used for irrigation based on actual use. In the areas with serious water depletion problems known as dark blocks, some incentives for keeping a part of land follow should be considered. Serious efforts should be made on supply side, to conserve maximum of rain water during the monsoon season. Incentives should also be provided to the farmers for installation of water saving devices like drips and sprinkler etc.

Since forest denudation is one of the most serious factors promoting the sustainability of both nutrition and food security, the need of the hour is to initiate a forest policy in all States of India. For this purpose, plants shall have to be selected not only on the basis of their fast growth but also from the point of view of their effects on agricultural development and environmental upgradation. Strict measures shall have to be taken to check there excessive and reckless exploitation of the existing forest.

For sustainable agricultural development, the inputs i.e. the seeds, fertilisers, pesticides and agricultural implements have to be combined in such a way as would meet the present challenges and would also ensure stability in the future. Choice of a suitable cropping is also very important in this direction. For ecological sustainability, other important measures like integrated use of renewable and non-renewable resources of energy, the integrated nutrient supply system including crop rotation and application of green manures and pest management including the conservation and wise use of genetic diversity need to be taken up in an appropriate manner.

Agricultural diversification through dairy farming or pulse crops are some of the options which can check soil degradation and restore soil health and fertility of degraded lands. Expansion of dairy would be useful in various ways. First it would result in shift in some area from wheat and paddy to fodder crops, some of which are legume. Secondly it would increase availability of organic manures i.e. animal dung for agricultural production.

Burning of crop residues and application of insecticides are the main sources of environmental pollution. Burning of crop residues of paddy and wheat produces smoke which makes the atmosphere suffocations in the months of April and October. Air pollution caused by the burning of crop residues can be checked by enhancing industrial demand for straws as it can be used to manufacture products like paper, mats and ropes etc.

In this volume of the book, eleven papers have been included. These articles emphasise on sustainability of agricultural development, farm sector reorganisation for sustainable agricultural growth and its optimal productivity, possibilities of substitution and complementarity between inputs, economies of scale in agricultural sector and the much debatable issue regarding the relationship between land productivity and firm size. Some of the articles also deal with the linkages of agricultural sector with rest of the economy, structural changes in the agricultural sector, effect of mechanisation on cropping intensity and district agricultural plans formulation.

C.H. Hanumantha Rao in his article discusses the growing importance of yield increasing technology in agriculture. He points out that the new seed-fertiliser technology has raised the response of output to water. Thus for a given variability in rainfall, the instability in output would be greater. However when new technology is applied under assured conditions, the increase in output would be on a stable path. The author points out that in the early phase of green revolution, inverse relationship between farm size and output per acre has disappeared because large farmers, owing to better access to capital resources stepped up yield per acre at a faster rate than small farmers. However, in the later phase of green revolution, due to increase in the supply of institutional credit to small farmers and relatively higher cropping intensity among small farmers, the inverse relationship between farm size and output per net operated acre started reappearing. Author further argues that the uneven regional growth was mainly responsible for low absorption of labour within agriculture. In the regions where there was abundant availability of labour, the growth in output was too slow to generate adequate employment opportunities. In high growth regions — labour was not abundant, and the sudden rise in the demand for labour, induced mechanisation and labour saving practices in general. The net result of it was a significant decline in the elasticity of employment with respect to agricultural output for the country as a whole.

Due to rapidly increasing population and increase in per capita income, maintaining rapid increase in productivity of foodgrains, especially wheat, is a matter of concern to the policy makers. D.S. Sidhu and Derek Byerlee in their article has examined this issue in detail. Their findings reveal that although the green revolution was widely associated with an increase in employment, the use of labour input per hectare has fallen during the post-green revolution period i.e. during 1966-75. Stagnant real wages during the above mentioned period are consistent with the fall in labour input use per hectare, as a result of mechanisation coupled with net emigration of labour force from other states to rural Punjab. According to Sidhu and Byerlee, during 1960s the farmers were the main beneficiaries from the increased productivity of wheat but in 1970's and 1980's consumers of wheat have received the bulk of benefit. The author suggests that further increase in the productivity of wheat crop will have to be achieved by more efficient use of inputs like fertilisers and water.

In earlier studies, researcher have found the relationship between land productivity and farm size to be inverse. But the recent evidence suggests that there exists no relationship between productivity and scale. These two contrary pieces of evidence have generated a controversy regarding the relationship between productivity and scale. Renu Kallianpur and Romesh Diwan, in their paper entitled, "Productivity Growth : Scale and Technology Analysis of Punjab Wheat Farm", reconcile this seemingly contrary evidence. The authors argue that the inverse relationship is consistent with the evidence of no relationship if scale-bias of technological inputs is taken into consideration. If biological technology is biased in favour of large farms, and the land productivity for small and large farms may not be different, then the reconciliation of seemingly contrary evidence lies in the scale biasedness of biological technology. Empirical results also show that the use of biological technology is biased in favour of large farms.

If the same quantity of technological inputs are used on both small and large farms and small farms give higher productivity per hectare, then the inverse relationship between farm size and productivity can be said to hold true. Also, if the output per hectare turns out to be same on large and small farms, even when the technological inputs used more on large farms, then one can conclude that inverse relationship between land productivity and farm size holds true. The empirical results also reveal that output per hectare on wheat farms of both large and small size does not differ significantly when large farmers use relatively larger technological inputs such as fertilisers, insecticides and pesticides etc.

Bina Agarwal has studied the impact of tractors and tubewells on cropping intensity in the Indian Punjab. She points out that tubewell irrigation seems to promote cropping intensity by making available an assured water supply throughout the year and by enabling closer supervision over the quantum and timing of irrigation. Tractors tend to add to cropping intensity because they can prepare the fields for sowing next crop in a comparatively shorter period. For all farms taken together, negative relationship between farm size and cropping intensity was observed. The author also pinpoints that with the use of tractors in place of bullocks, much saving in fodder land cannot be expected because only a small part of bullock feed come from fodder and the bulk of it comes from wheat and rice straws.

Kanchan Chopra has empirically examined the issue of substitution and complementarity between inputs like water, land, fertilisers, labour and capital. Her results show that there is a considerable degree of complementarity between land, labour and to some extent conventional capital. The availability of irrigation facilities and mechanical equipments could encourage more intensive use of land. Further, she points out that paddy cultivation present considerable scope for substitution between inputs. There is much scope for reproducible resources such as machinery and fertiliser to replace exhaustible resources like land and water. Such substitution will, of course have a cost attached to it particularly as machinery and fertiliser are both foreign exchange using inputs. The study also shows that the use of tractors is not significantly different across different size-classes of farms partly because of the large scale presence of a market for tractor services. Either tractor use or intensity of land use is not related to the farm size. This is in conformity with recent contention that with the introduction of High Yielding Varieties of seeds, fertilisers and tractors, the larger farmers have caught up with the smaller ones in the intensity of land use and the much discussed inverse relationship between farm size and intensity of land use no longer holds true.

Robert Evenson has empirically examined the question of the Green Revolution's effect on North Indian agriculture. For this analysis, the author has used an approach based on system of output supply and factor demand equations derived from a maximised variable profit function. The results show that the green revolution in India has been decisively non-neutral in its effect on both output supply and factor demand. It has favoured the supply of wheat and rice at the expense of other crops on the output side. It has also favoured the demand for fertiliser, tractors and pesticides etc. at the expense of bullocks and human labour on the input side.

G.K. Chadha and R.K.Sharma in their paper examine empirically the relationship between farm size and cropping intensity and farm size and irrigation base, using 1970-71 agricultural Census Reports data for various states. They find an inverse relationship between farm size and the proportion of net sown area irrigated. The main reason for this, according to the authors is that the small farmers, in the absence of tubewell irrigation, could irrigate a large proportion of their net sown area. The authors conclude that despite positive relationship between farm size and improved irrigation base, there exists a negative relationship between farm size and cropping intensity in a majority of districts. This means that inspite of a technological disadvantage in the matter of improved irrigation facilities, the marginal and small farmers continue to cultivate their lands more intensively over a large part of Indian agriculture.

K.S.Dhindsa in his paper analyses structural changes in Punjab agriculture. According to him, structural changes in agriculture include changes in the resource structure of agriculture, distribution of land holdings, changes in the composition of workers engaged in agriculture and percentage changes in contribution to the state's Net Domestic Product of different components of the primary sector. For estimating the effect of traditional and modern resource input on agricultural output, the Cobb-Douglas production function has been used. The author points out that there has been an increase in the productivity per hectare of almost all the crops. But the increase in productivity of foodgrain crops especially wheat, maize and bajra has been greater than that of non-food crops. The author concludes that such a remarkable increase in the productivity per hectare of foodgrain crops has been made possible by the bio-irrigo-chemical technological breakthrough brought about by the scientists and the responsive farm community. The traditional capital goods represented by draft animal, ploughs and carts have lost their importance in Punjab's agriculture. The author also points out that there has been a significant increase in the number of agricultural labourers and this confirms the belief that the process of proletarianisation is growing at a rapid pace in the rural areas of Punjab.

It is observed that agriculture's dependence on itself declines and that on other sectors of the economy increases as the nation's economy develops over time. It seems, therefore desirable to study the process overtime to examine whether the agricultural sector of an economy is independent or inter-linked with rest of the economy. S.Suryaprakash, M.V.Srinivasa Gowda and J.V.Venkataram in their paper, "Relative Linkages of Indian Agriculture", have pointed out that the sector-wide and economy-wide linkages of different sectors of an economy are helpful in analysing the production structure of an economy. This will help in arriving at suitable strategies for economic development. But these are not helpful directly in sectoral planning, more so when the sector remains isolated from rest of the economy. In such cases, relative linkages is a more useful measure.

The authors argue that the agricultural sector in India has shown a tendency to rely less on itself for inputs (in a relative sense). This is in confirmity with the general pattern observed world-wide that with development, sectors open up more and increase their dependence on the economy. But, this may not necessarily be a welcome feature as the present thinking is diverted more towards organic farming and sustainable agriculture which clearly imply reductions in off-farm inputs use and increase in the dependence on own inputs. It is also noticed that the economy-wide linkages of a sub-sector change over time, partly due to changes in growth rates of sub-sectors' output. Thus, it may be said that the on input-side relative linkages are indirectly influenced by the output growth rate as well.

This comprehensive section deals with articles on sustainability of agricultural development, farm sector reorganisation for sustainable agricultural growth and its optimal productivity, possibilities of substitution and complementarity between inputs, economies of scale in agriculture sector and the much debatable issue regarding the relationship between land productivity and farm size. Other articles deal with linkages of agricultural sector with rest of the economy, structural changes in the agricultural sector, effect of mechanisation on cropping intensity and district agricultural plans formulation.

In his paper, "Achieving Sustainability and Extending the Green Revolution", M.S.Swaminathan points out that the decline in environmental quality has underlined the need for harmonising the logic of economics with that of ecology. There is an urgent need to consider methods for achieving continuous advances in biological productivity in an ecologically sustainable manner. Industrialised countries have intensified their efforts to assert soil erosion, conserve water and biological diversity and reduce the consumption of mineral fertilisers and chemical pesticides and at the same time stabilise the per hectare productivity at the current high levels. For India and China and many other developing countries, the challenge is how to produce more from diminishing land and water resources. Therefore, such countries will have to develop appropriate technologies and public policies for achieving the desired rate of growth without being detrimental to the long term production potential of soil and water.

The author emphasises that farmers living in CDR (complex, diverse and risk-prone) areas will need technologies tailored to their agro-ecological and socio-economic circumstances. It is theoretically possible to breed high-yielding varieties of cereals and other crops possessing multiple resistance to a wide range of biotic and abiotic stresses. In many CDR areas, systems of land use may have to be altered to make them ecologically sound.

Swaminathan also points out that the productivity path way of production improvement is the only option available to countries like India and China for meeting their needs for agricultural commodities. An appropriate integration of modern and traditional technologies can help to combine the ecological and employment intensity strengths of traditional technologies with the economic efficiency characteristics of modern technologies. Genetic strains which are attracting more attention these days can fix nitrogen in stem, in addition to fixation in the rhizosphere are very much less dependent on market-purchased chemical inputs and green manure crops.

S.S.Johl in his paper, "Farm Sector Reorganisation for Sustainable Growth and Optimal Productivity" points out that in order to avoid food shortages and spiral in food prices, the government must check the growth of population and produce enough food products on a sustained basis. He also makes a distinction between the 'Need' and 'Demand' for food products. The people living below the poverty line have not enough purchasing power to buy sufficient quantity of foodgrains as required by them at the prevailing market prices. So India is surplus in foodgrains because of lack of enough purchasing power on the part of the poor people. He also suggests that changes in the food habits of people from wheat/rice towards potatoes and vegetables etc. can also solve the problem of food shortages.

Johl argues that since 78 per cent of the farmers have less than two hectares of land holdings and these farmers are incapable of adopting improved production technology, there is a pressing need to encourage co-operative farming in areas where intensive or high value crop cultivation can be pursued. For this purpose, land of adjoining farmers can be pooled to create farm corporations of 500 to 1000 hectares each in those areas where vegetables, flowers, fruit and dairy farming can be pursued with advantage. The management of these farms can be entrusted to democratically elected small board of managements and sub-committees that decide upon the production patterns, technology adoption, financing, capital investment as well as market development, sales, processing, storage of produce etc. Thus to meet the challenge of providing food security to the fast growing population on a sustained basis, production, processing, distribution and market developments have to proceed together in an integrated manner.

1

TECHNOLOGICAL CHANGE IN INDIAN AGRICULTURE

Emerging Trends and Perspectives

C.H. HANUMANTHA RAO

Let me, at the very outset, express my deep sense of gratitude to the members of the Indian Society of Agricultur al Economics for the honour done to me by electing me the President for this annual Conference in the Golden Jubilee Year of the Society. It is quite in the fitness of things that the Golden Jubilee Conference of the Society should be held in Bombay under the joint auspices of the Department of Economics of the University of Bombay and the Indira Gandhi Institute of Development Research, Goregaon (East), Bombay. The Society was born and nurtured in the precincts of the University of Bombay and the stalwarts in agricultural economics from this University have been guiding the affairs of the Society during half a century of its existence.

The very first time I attended the Annual Conference of the Society was in its Silver Jubilee Year in 1964 at Anand. The last 25 years have witnessed major changes in Indian agriculture. Undoubtedly, by far the most important among these changes is the introduction and the widespread use of new technology such as high-yielding varieties (HYVs) of seeds and fertilisers. This has helped us to achieve self-sufficiency in foodgrains and improve the productivity of resources in agriculture. Above all, our farmers are becoming increasingly aware of the benefits that can accrue from the application of science and technology to agriculture and have been responding favourably to the new practices recommended by the Extension Services.

We as professional economists have, by and large, reacted holding constructively to these developments in Indian agriculture. Apart from holding annual conferences, our Society, under the leadership of Professor M.L. Dantwala and Professor V.M. Dandekar, has been organising seminars on themes relevant to this phase of rapid technological transformation of Indian agriculture. The scholarly papers on such themes appearing in our Journal bear testimony to the deep interest shown by our profession in the processes of this transformation.

I propose in this address to reflect on some of the broader issues bearing on the technological changes in Indian agriculture, particularly in the light of the emerging trends in the recent period.

It was becoming clear by the mid-sixties that there was no alternative to technological change in agriculture for achieving self-sufficiency in foodgrains. Even those countries in Asia which could carry out radical land reforms and build up adequate infrastructure for agriculture had taken to the path of modernising agriculture. Japan took the lead in this direction and China followed suit even after successfully experimenting with structural changes and mobilisation of growing labour force for capital construction in agriculture.

After the mid-sixties, apart from the introduction of HYVs of seeds for wheat and rice, public investment in agriculture was stepped up significantly. The new technology raised the profitability of investment for the farmers. Besides, in agriculture, there is a high complementarity between public and private investment. As a result, the annual gross capital formation in agriculture, public and private combined, at constant prices, rose much faster than in the pre-green revolution period (see Fig. 1.1). This is reflected in the accelerated increase in gross irrigated area. From less than one million hectares per annum during the pre-green revolution period, it increased to about 2.5 million hectares per annum during the seventies. However, owing to the land constraint, the growth of net sown area slowed down considerably. This explains why despite the accelerated increase in capital formation and the introduction of new technologies, the growth rate of agricultural output was no higher than in the pre-green revolution period.

Nevertheless, there was a rapid increase in the marketed surpluses of foodgrains and in the stocks held by the government through internal procurement. This is due to the basic changes in the structure of agricultural production witnessed in this period. First, the breakthrough in foodgrains output occurred in the developed regions in the North-West and in certain parts of the South. The consumption of foodgrains was already high in these areas and therefore a considerably

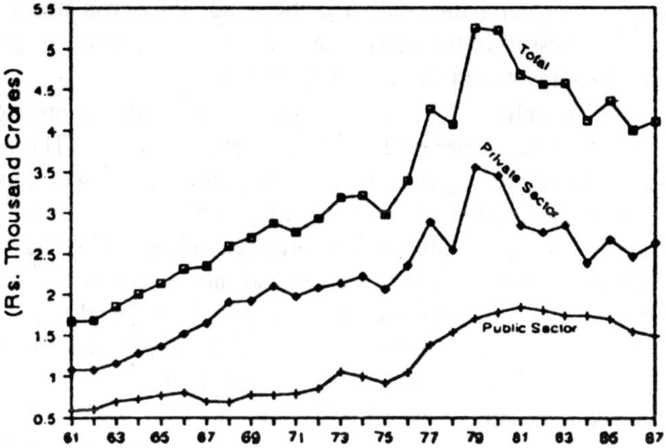

Fig. 1.1 : Gross Capital Formation in Agriculture (1980-81 Prices)

Source : Government of India (1989).

larger proportion of additions to output was marketed. Second, the use of labour per unit of output showed a significant decline, which also resulted in the generation of larger foodgrain surpluses from the rural areas.

There are two other factors responsible for the rise in marketed surpluses and availability of foodgrains. The increase in output during this period has been taking place mainly through the increase in yield per hectare. Therefore, there has been a decline in the ratio of seed to output, although our official estimates continue to be based on the conventional norms. Secondly, the rate of urbanisation has increased and, for various reasons which I need not discuss here, the average per capita consumption of foodgrains in the urban areas has been significantly lower than in the rural areas.

We are used to comparing the performance in the pre-green revolution period with that in the post-green revolution period as a whole. However, the mid or the late seventies mark a definite departure from the trends observed in the early phase of green revolution in several respects. I would, therefore, like to draw your attention to some of the emerging trends in the recent period.

There seems to be some improvement in the growth rate of foodgrains output in the last decade, i.e., 1978-79 to 1988-89, when compared with the first decade of Green Revolution. Crops such as rice, pulses and oilseeds, whose growth rates in the first decade of green revolution had fallen much below those recorded in the pre-green revolution period, are now showing higher growth rates (Table 1.1).

There are thus clear indications that the major inter-crop imbalances in growth witnessed in the early years of green revolution are getting redressed to some extent in the recent period.

Oilseeds and pulses are grown essentially in the rainfed or unirrigated areas. Even for rice, nearly 60 per cent of area is rainfed. The recent experience, therefore, suggests that the disparities in growth between the irrigated and rainfed or dry areas may not be as sharp as in the early years of Green Revolution. It is also heartening to find that many of the States where poverty is widespread and where the growth of foodgrains output had slowed down in the first decade of Green Revolution, e.g., Assam, Bihar, Orissa, Madhya Pradesh and West Bengal have shown a much better performance in the last decade. Their performance is nearer, or even higher than, the all-India average. The growth of rice had slowed down considerably in the Eastern States during the seventies. All these States have experienced a step-up in the growth of rice during the last decade and some of them have recorded higher growth than the national average (Table 1.1).

TABLE 1.1 : Statewise Compound Growth Rates of Output

States	Rice		Foodgrains	
	1967-68 to 1977-78	*1978-79 to 1988-89**	*1967-68 to 1977-78*	*1977-78 to 1988-89**
(1)	(2)	(3)	(4)	(5)
Andhra Pradesh	2.66	1.70	2.56	0.41
Assam	0.88	2.12	1.27	1.63
Bihar	0.82	2.83	1.14	3.52
Gujarat	3.39	–0.43	2.23	–4.97
Haryana	11.34	5.12	2.86	3.94
Himachal Pradesh	0.04	–1.14	1.24	0.36
Jammu & Kashmir	1.73	0.09	1.47	0.36
Kranataka	–0.01	-0.44	2.50	–0.58
Kerala	0.54	–2.36	0.48	–2.30
Madhya Pradesh	0.96	5.30	1.23	4.36
Maharashtra	4.79	–0.44	4.84	–0.46
Orissa	–1.00	2.61	0.80	2.59
Punjab	18.21	7.04	5.29	4.57
Rajasthan	10.90	–1.77	3.03	1.37
Tamil Nadu	1.45	0.89	1.69	0.74
Uttar Pradesh	4.06	7.32	4.06	5.16
West Bengal	1.71	5.23	2.60	4.53
All-India	2.16	3.19	2.31	2.68
Pulses	0.27	2.32		
Oilseeds	1.39	3.86		

*Estimates for 1988-89 are provisional.

Source: Directorate of Economics and Statistics, Ministry of Food and Agriculture, Government of India, New Delhi.

This encouraging performance is basically attributable to the spread of new technology to the lagging crops and regions. The special programmes for rice and oilseeds and increasing attention being given to dry farming and to the less developed States have started yielding results. This is corroborated by the increase in the consumption of fertilisers during the eighties at a faster rate in the Eastern and Western regions than in the Northern and Southern regions. What is more, fertilisers consumed in *kharif* season increased at a much faster rate than in *rabi* season, particularly in the Eastern and Western regions, indicating the spread of new technology to the rainfed and environmentally unfavourable areas. There is also evidence of an increase in the share of the Eastern States in the total gross irrigated area of the country.

However, the picture is not encouraging regarding capital formation in agriculture in the country during the eighties. After showing a steep rise in the seventies, the gross capital formation in agriculture at constant prices, both by the public and private sectors, showed a significant decline in the eighties (Fig. 1.1). The decline in the net capital formation could be even steeper. While the gross irrigated area increased at the rate of nearly 2.5 million hectares per annum during the seventies, the increase in the eighties came down to less than a million hectares per annum, on an average.

The gains in foodgrains output in the last decade have come essentially from the improved utilisation of the available infrastructure and from the increase in yield per hectare. Already, during the first decade of Green Revolution increase in yield per hectare accounted for as much as 80 per cent of the increase in foodgrains output. During the last decade, there was some decline in the area under foodgrains. The entire output growth in this period is, therefore, attributable to increase in yield per hectare. This underscores the growing importance of yield-increasing technology.

Now, one can legitimately ask the question : At what cost has this growth been achieved? Indeed, based on the input data from the National Accounts Statistics, fears have been expressed that growth in agricultural output is being achieved with increasing costs per unit of output, as the index of inputs at constant prices has been rising faster than output.[1] However, these data on inputs exclude land and labour. The new agricultural technology is essentially land-saving and also labour-saving to some extent. There is substitution of capital for land and labour. It is not surprising, therefore, that with the increasing application of new technology, the capital inputs should rise faster than output. What we should really be looking for is the productivity of total inputs-land, labour and capital combined.

The estimates of costs of production brought out by the Ministry of Food and Agriculture, Government of India give the latest position in regard to the shares of land, labour and capital in the total costs. Apply these weights to the growth of different inputs, the index of total inputs is seen to be rising at a little less than 2 per cent per annum in the post-green revolution period against the growth in foodgrains output at 2.6 per cent per annum. On this basis, between one-fifth to one-fourth of the observed increase in the output of foodgrains can be attributed to the increase in the productivity of inputs. Commodity-wise, the contribution to the growth of output from the improved productivity of inputs may be nearly 45 per cent for wheat and about 15 per cent for rice (Tyagi, 1989).

Although the growth rate in foodgrains output seems to have improved because of the impact of new technology, the growth performance has by no means been smooth. The instability or the year-to-year variation in foodgrains output has increased in the post-green revolution period. Even within the post-green revolution period, the instability in the last decade (1978-79 to 1988-89) was higher than in the preceding decade. The rural poor, particularly in the drought-prone areas and in the remote areas of the country, continue to suffer from fluctuations in employment and income and inadequate availability of foodgrains in years of drought.

However, this increase in instability cannot be attributed to new technology. Rather, the instability arises from the adverse agro-climatic conditions in which the technology is used. The new seed-fertiliser technology has raised the response of output to water. Thus, for a given variability in rainfall or moisture conditions, the instability in output would be greater. However, when the new technology is applied under assured irrigated conditions, the increase in output would be on a stable path. This is exactly what has happened in the case of wheat where the new technology has made the maximum impact. The rise in instability is significant for crops like rice, oilseeds and pulses. This is explained by the increasing application of new seed-fertiliser technology for these crops in rainfed areas and under uncertain irrigation in the recent period. This suggests the need for stabilising consumption through larger procurement of foodgrains when harvests are good and their liberal release through public distribution system for the vulnerable groups in years of drought.

Perhaps the most widely debated issue about Green Revolution is the growing disparities in income between different regions and classes of farmers. The experience after the mid-seventies covering nearly a decade and half reveals trends which are typical of a diffusion process. These trends are; the spread of green revolution to new areas, the increasing adoption of new technology by the small farmers, the decline in the relative prices of foodgrains (see Fig. 1.2) and the rise in real wages in agriculture in the less developed regions where new technology is spreading.

Even so, the expectation that land-saving technological changes in a labour-abundant economy like India would lead to greater labour absorption in agriculture has been largely belied. The short-duration HYVs of seeds can facilitate an increase in cropping intensity and may thus lead to an increase in labour use. In practice, however, because of the profitability of intensive farming involving the use of large quantity of water and fertilisers per unit of land, there was not much incentive for increasing cropping intensity. Moreover, until recently, the impact of new technology and the expansion of tubewell irrigation was not significant in regions where labour is plentiful. This may explain why the cropping intensity has risen at a much slower rate in the post-green revolution period.

In the early phase of Green Revolution, large farmers, owing to better access to capital resources, stepped up yields per acre at a faster rate than small farmers. Because of this, in areas experiencing technological change, the inverse relationship between farm size and output per acre began to disappear. In course of time, however, the supply of institutional credit for the less developed regions and small farmers improved significantly (Desai, 1988; Haque and Verma, 1988). As a result of this and also because of improved extension services, the use of new seed-fertiliser technology among small farms caught up with that among large farms. And, because of the continued advantage that small farmers have in respect of cropping intensity, the inverse relationship between farm size and output per net operated acre has started reappearing (Bhalla and Chadha, 1983; Shergill, 1987). However, labour input per acre continued to decline sharply with the increase in the size of holding.

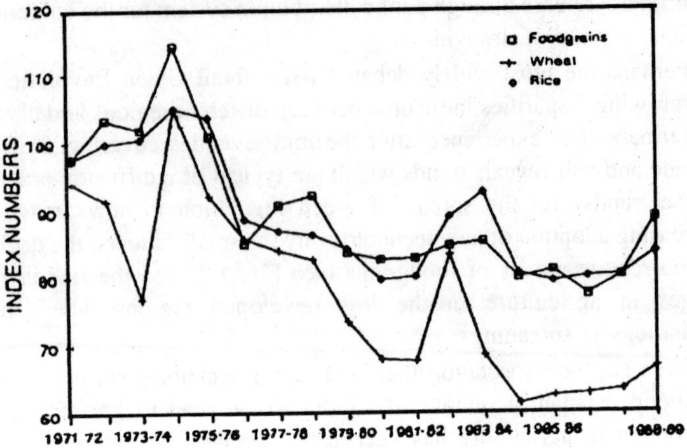

Fig. 1.2 : Index Number of Relative Prices
(1970-71 = 100; Deflated by Wholesale Price Index)

Source : Commission for Agricultural Costs and Prices, Government of India.

Insofar as large farmers have resumed land for self-cultivation from share-croppers because of the profitability of new technology, it has led to lower labour absorption and reduced incomes for erstwhile tenants. However, in quite a few places, cost-sharing on new inputs between landowners and share-croppers has become popular resulting in the sharing of benefits from new technology by the landowners as well as tenants.

Even in the early phase of Green Revolution, the emerging disparities between different regions were more conspicuous than the disparities between different classes of farmers within regions experiencing technological change. The uneven regional growth was mainly responsible for low absorption of labour within agriculture. In a large number of States, especially in those regions where there was abundant availability of labour, the growth of output was too slow to generate adequate employment opportunities. In high growth regions labour was not plentiful and wage rates were high. The sudden rise in the demand for labour in these areas induced mechanisation and labour-saving practices in general. This happened despite the use of migrant labour from the less developed regions for certain operations. The net result was a significant decline in the elasticity of employment with respect to agricultural output for the country as a whole (Rao, 1975; Bhalla, 1987; Vaidyanathan, 1986). At the same time, the Inter-state disparities in the growth of output and in output per worker continued to increase from the early seventies through the early eighties (Bhalla and Tyagi, 1989).

However, the Inter-State disparities in agricultural wages started declining since the mid-seventies (Jose, 1988; Nayyar, 1987). The decline in disparities in real wages is even more significant. This is because the decline in the relative prices of foodgrains had a greater impact on the purchasing power of wage earners in low wage areas. For example, the disparities in real wages in agriculture between the Punjab, on the one hand and Bihar, on the other, started coming down since the mid-seventies (Fig. 1.3). The real wages in the Punjab having risen significantly in the early phase of green revolution, virtually stagnated after the mid-seventies whereas the real wages in Bihar started rising. The causes for the decline in inter-State disparities in real wages are: the out-migration of labour from the less developed regions to the high wage areas; mechanisation of agricultural operations in the developed regions; the decline in the relative prices of foodgrains; the increase in employment generated in the less developed regions under the poverty alleviation programmes; and the recent pick up in agricultural growth in these regions. An encouraging consequence of these developments is that the disparity between male-female wage rate has started declining in several States. The use of new technology may have improved the bargaining power of female labour for operations such as transplanting and interculturing on account of the rise in seasonal demand for such labour.

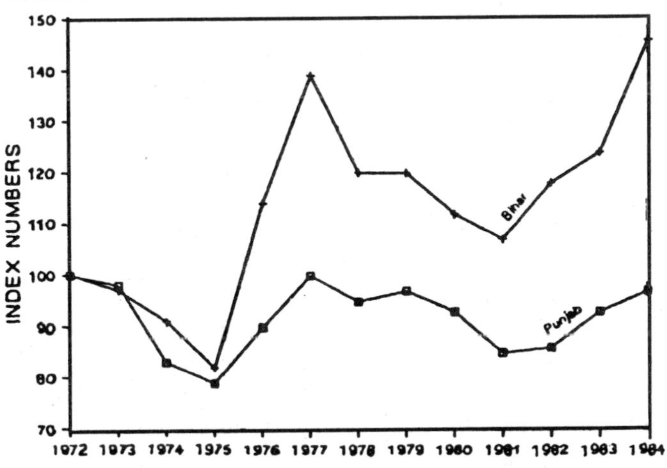

Fig. 1.3 : Index Number of Real Wages (1970-71 = 100)

Source : Jose (1988).

Low income groups, both in the rural and urban areas, would derive greater benefit from the decline in the relative prices of foodgrains, because they spend a larger proportion of their income on foodgrains than the upper income groups. Agricultural price policy has been able to ensure that the productivity gains from technological change are shared both by the consumers and producers. However, consumers seem to have benefited relatively more than the producers from the productivity gains, particularly in the case of rice and jowar. The demand for wheat has been more favourable than for other cereals. So the decline in the relative price of wheat was only about 25 per cent, while the productivity gains amounted to nearly 45 per cent of output growth.

The indirect contribution of Green Revolution to equity through food security and the decline in the relative prices of foodgrains has perhaps been more significant than its direct impact by way of labour absorption in agriculture. Because of slow growth of employment in agriculture, the employment and income generated under poverty alleviation programmes had to be stepped up by using surplus stocks of foodgrains. From about 20 million man-days of employment generated annually in the mid-sixties, the employment generated under such programmes in the country as a whole amounted to 850 million man-days in 1988-89. These employment programmes together with the income generated under the Integrated Rural Development Programme (IRDP) seem to make up for about half the deficiency in employment generation in agriculture in the post-green revolution period. The relative contribution of these poverty alleviation programmes to employment generation is more significant in the poorer regions where agricultural growth has been below the national average. These programmes have been made possible because of the increased availability of foodgrains from internal procurement.

Pollution of environment due to the intensive use of chemical fertilisers and pesticides has become a major problem in the developed countries. In India, environmental degradation in the rural areas has arisen not so much from the high level of chemical inputs used as from deforestation and extension of cultivation to ecologically fragile areas. Land-saving technological changes by reducing pressure for extension of cultivation and by augmenting bio-mass, contribute to the conservation of fragile areas and regeneration of forests.

Across different States in India, the extension of area under cultivation and the denudation of forests seem to be high where the progress of yield-increasing technology is slow. In such regions, the levels of agricultural income and wages are low and poverty levels are high. Similarly, the pressure from animals such as goats and sheep on forests and common lands has been increasing in regions where growth in crop production is slow. This is because the rural poor supplement their incomes by rearing these animals (Rao, forthcoming; Mellor, 1988). Apart from yield-increasing technology for crops, the use of new technologies in forestry and animal husbandry to raise the yields from trees and animals and to improve management practices are likely to contribute significantly to the protection of environment. The emerging biotechnologies are opening up opportunities for environmentally sound management of agriculture. I shall come back to this aspect a little later.

So far I have discussed the recent experience relating to technological changes in Indian agriculture in the light of the objectives of self-reliance, productivity, stability, equity and sustainability. I would now like to discuss briefly what I consider to be the major issues in technology strategy and policy and the perspectives before us.

The presumption that green revolution technologies have been basically evolved within the developing countries in response to the prevailing factor endowments is not an accurate description of the reality. Many of them have been, in a large measure, transferred from the developed country situations and adapted to the specific circumstances of the developing countries. Therefore, it becomes necessary to understand the feature of these technologies which are influenced by the circumstances of the developed countries. It is also necessary to understand the circumstances in the developing economies responsible for their transfer and congenial for their adaptation.

Several developed countries, notably the U.S.A. and Japan, experienced major break-throughs in crop yields during the first two decades following the Second World War. The infrastructures for agriculture, including water management in particular, were very well developed in these countries. The new seed-fertiliser technologies were evolved there to suit the favourable or relatively homogeneous agro-climatic environments. Moreover, agricultural, breakthrough in this period followed the industrial take-off. Apart from the availability of industrial inputs, labour was becoming costly for agriculture. This happened in such diverse situations as the U.S.A. where land was not a scarce factor and in Japan where there was an extreme scarcity of land. The new bio-chemical technologies were, therefore, not only land-saving but also labour-saving and highly capital-using. Further, the entire institutional framework including farmers' consciousness and administrative capabilities were conducive to the speedy absorption of new technologies. This explains why the expenditure on extension services as a ratio to the expenditure on agricultural research was very low in these countries when compared with the developing countries.

The challenge of growing population pressure, in the post-war period, on the limited land available was compelling the governments in many of the developing countries to borrow the yield-increasing technologies from the developed economies. It is not surprising, therefore, the land-saving technology was widely adopted in India in the earl phase of green revolution in areas where the availability of land per worker was significantly higher than the national average. One the other hand, regions where land was extremely scarce and labour was abundant lagged behind for quite some time in the adoption of land-saving techniques. This is contrary to what the theory of induced innovation would lead us to expect (Hayami and Ruttan, 1985; Rao, 1976; Janvry and Dethier, 1985; Grabowski, 1979, 1981; Hayami, 1981; Pinstrup-Andersen, 1982).

Environmental or agro-climatic suitability of new technology is, to my mind, the most important factor explaining the regional variation in the impact of new technology. It is understandable that the areas endowed with favourable environment should also be resourceful both in terms of investible capacity and institutional preparedness. The regions with low to medium rainfall but served with assured sources of irrigation have proved so far to be most suitable for the profitable application of new seed-fertiliser technologies. These areas are sufficiently exposed to sunshine and are relatively free from the incidence of pests and diseases. It is also not accidental that the top four States in the country in respect of per capita income have taken a lead in the adoption of new technologies. These are the Punjab and Haryana which have assured irrigation facilities and, Gujarat and Maharashtra which are mainly dependent or rainfall. Apart from the resourcefulness of farmers, these State Government have adequate resources and have relatively efficient and responsive administration.

The role of well-endowed farmers in influencing the priorities in agricultural research has been well recognised. However, the governments in a democratic polity are also subject to pressures from the consumers and the majority of farmers who are resource-poor. The pressures from these groups can prove to be decisive in the long-run in determining the course of agricultural research.

Economists, true to their analytical tradition of treating technological change as an exogenous variable, tended to regard the available seed-fertiliser technology as unalterable. A large body of economists thought that socio-economic policies have to be designed to realise the full benefits from the given technologies under the given environments. On the other hand, quite a few felt that there has to be modification of environments through infrastructural development and reform of institutions before new technologies are introduced.

However, agricultural scientists who engineer the technologies cannot treat the available technology as given and unalterable. While the economists were still debating on the socio-economic consequences of the introduction of new technology, the scientists at the International Agricultural Research Centres with rare consensus decided to change the focus of research. They were striving to evolve technologies suited to unfavourable agro-climatic environments, and packages and practices to suit the resource-poor farmers. In India, the view that there was a need to change the environment to fit the technology as well as to develop technology to fit the environment was gaining strength (ICAR and IRRI, 1977; IRRI, 1982)

The new priorities in agricultural research in India were—(i) to reduce yield variability in rainfed areas by evolving High Stability Varieties (HSV) rather than introducing merely high-yielding varieties (HYVs). This requires varieties with multiple resistance to pests and droughts as also conservation of soil moisture; (ii) to minimise costs to the farmers through various input-saving practices including biological fixation of nitrogen; and (iii) to evolve cropping and farming systems to raise the incomes of farmers by making the optimum use of their resources throughout the year.

These new priorities require greater interaction with field situations through location—specific research, as unfavourable environments are highly heterogeneous. Besides, greater participation by the farmers in testing out research becomes necessary. Greater collaboration between agricultural scientists and social scientists, particularly agricultural economists, is also needed for understanding the socio-economic rationale of the prevailing systems. Such a knowledge would help to tailor technology to suit the resource endowments and needs of the farmers.

The yield increases in the last decade for lagging crops like rice, oilseeds and pulses, have come more from the vigorous extension of available technology to new areas through special programmes than from breakthroughs in technology. This experience underscores the importance of special programmes and input subsidies in a situation where the Green Revolution of the early period has brought down the relative prices of output, where yield variability is high and where farmers are resource-poor and risk-averse. Special efforts to ensure remunerative prices to farmers, particularly for oilseeds, also had a significant impact. The impact of technology as such on yields seems to be more visible in the case of rice.

The results of National Demonstrations conducted on farmers' fields under the Transfer of Technology Projects of the Indian Council of Agricultural Research, suggest a large gap between the demonstration and national average yields and that the yields can be increased 2 to 3 times in all the major crops. However, under similar agro-climatic conditions, the yield gap between the actual and the potential is, in general, no more than 50 to 60 per cent (Prasad Choudhary and Nayar, 1987). The yield gap between the actual and the potential seems to be higher under rainfed conditions and during the kharif season, suggesting that the input use by farmers may be restricted owing to uncertain rainfall conditions (Table 1.2). Since these yield demonstrations are individual crop-oriented, it is not known how well these new varieties would fit into the cropping systems of the farmers. The deficiency is most glaring in respect of cropping and farming systems research which is perhaps the most difficult area, as it is both multi-disciplinary and field-oriented.

The net returns per rupee invested seem to be higher under irrigated conditions or for rainfed *kharif* crops when soil moisture is adequate. This potential for raising yields and reducing their variability has prompted the integration of the task for genetic improvements in seeds with the measures for soil and moisture conservation in the dry farming strategy (Jodha, 1989). The National Watershed Development Programme which has been designed for rainfed areas with water-harvesting as one of its major components, seems to have taken a shape after years of experimentation and can be expected to yield significant results in the years to come.

TABLE 1.2 : Results of Demonstrations Conducted by ICAR Transfer of Technology Projects

Regions	*Kharif 1987*		*Rabi 1988*	
	Per cent difference between demonstration and control yield	*Net return per rupee of operating cost*	*Per cent difference between demon-stration and control yield*	*Net return per rupee of operating cost*
(1)	(2)	(3)	(4)	(5)
Rainfed				
Groundnut	53	1.23	—	—
Soyabean :	56	1.68	—	—
(Madhya Pradesh)	42	2.00	—	—
Sesamum	61	1.39	54	0.74
Sunflower	63	1.20	38	1.13
Mustard	—	—	100	2.19
Safflower	—	—	68	1.38
Irrigated				
Groundnut	40	1.33	34	1.77
Sunflower	70	1.86	27	1.62
Paddy (Southern States, 1988)	102	2.14	89	2.14
Mustard	—	—	55	1.77
Sesamum (Tamil Nadu)	—	—	28	1.29
Ragi (Southern States)	—	—	119	1.56

Source : Computed from data contained in the Reports of the Indian Council of Agricultural Research (ICAR), New Delhi.

The achievements of the Indian Agricultural Research System so far in respect of raising yields and reducing variability in the unfavourable agro-climatic regions are not comparable with those realised for the favourable environments in the early years of green revolution. The tools of emerging bio-technology like genetic engineering, tissue culture, etc., offer significant possibilities for breaking these barriers. Apart from possibilities for raising yields and improving the nutritive content of the products, bio-technologies will release, to a significant extent, the agro-climatic constraints to crop production by genetically building up resistance in seeds to moisture-stress, pests an diseases and adverse soil conditions. Through these properties and also through biological fixation of nitrogen, costs on chemical inputs can be reduced significantly. Bio-technologies may contribute significantly to the regeneration of degraded environment, as they help to release marginal lands from cultivation of crops, restrict the use of chemical inputs and facilitate afforestation through tissue culture techniques.

Bio-technologies present a wide range of options. Whereas they can potentially be harnessed to suit the difficult environments and poor farmers, they can also be engineered to suit the entrenched interests to the detriment of the poor. The classic example is that of the choice between pest-resistant versus pesticide-resistant varieties of seeds. The former can be pro-poor as well as environmentally sound because of the savings on pesticides, whereas the latter can promote the interests of multinationals doing business in pesticides (RIS, 1988).

The high flexibility of bio-technology enhances the role of research strategy and policy. The role of governments of developing countries for setting up right priorities in bio-technology research and for achieving self-reliance in R&D becomes even more essential than in the case of green revolution technology. Since favourable environments (controlled irrigation under low to medium rainfall conditions) may not account for more than one-third of cultivated area in India even after the full potential for irrigation is realised, it may be desirable, from the point of view of growth as well as equity, to invest in the generation of new technologies for the unfavourable area (David, Otsuka and Hayami, 1989). The possibilities opened up by new bio-technology raise the profitability of such investment in agricultural research.

Bio-technologies will save significantly on conventional resources—land, labour and capital. They will be knowledge-intensive and skill-intensive. Therefore, the distribution of gains from bio-technology will depend increasingly on the access to new inputs and knowledge, on agriculture-industry relationship and on the dependence of domestic agriculture on Transnational Corporations for critical inputs and services. Public intervention to strengthen the human capabilities of farmers, for widening the base of agro-services and for counteracting the restricitive practices of Multinational Corporations is going to be critical for the speedy adoption to bio-technologies as well as for equitable sharing of benefits. The direct contribution of agriculture toward labour absorption is likely to be even more limited than in the case of green revolution. Food-population imbalance is certain to be overcome but access to food through the generation of employment and purchasing power has to become a major concern of development planning in general.

Because of the wide range of options opened up by bio-technologies, there is a greater scope for economists and social scientists to contribute towards the evolution of appropriate technologies (Swaminathan, 1989; Raman, Balaguru and Samanta, 1989). Economists have been active, particularly in the early years of green revolution, in doing ex-post analyses relating to the profitability of new input packages, factors influencing the adoption of new technology and how the gains from technology are distributed. Our professional contribution towards the evolution of appropriate technologies, either in collaboration with agricultural scientists or at the policy-making level, has been limited. May I, therefore, suggest that, in its Golden Jubilee Year, the Indian Society of Agricultural Economics may initiate collaborative ventures with the Indian Council of Agricultural Economics may initiate collaborative ventures with the Indian Council of Agricultural Research to bring together agricultural economists and agricultural scientists for evolution of appropriate technologies and for shaping agricultural technology policy.

I have chosen to discuss in my address mainly the issues concerning technology. Needless to say, the declining trend in real public investment in agriculture has to be reversed, if the potential for agricultural growth is not to come down. Among different sector, public investment in irrigation and flood control showed the biggest shortfall in real terms when compared with the outlays contemplated both in the Sixth and Seventh Plans. States particularly lagging behind are the less developed ones that are short of resources such as Uttar Pradesh, Bihar, Madhya Pradesh and Orissa.

Special programmes and subsidies for lagging regions and small farmers would serve to neutralise the disincentive effects of falling relative prices of output. At the same time, such a policy would ensure the flow of benefits to the consumers from the increase in productivity. Improvement in overall institutional framework for agriculture, including consolidation of holdings and strengthening credit, marketing extension and input-delivery systems is essential for speedy absorption of new technology and for equitable distribution of gains.

Note : Reproduced from *Ind. Jn. of Agri. Econ.*, Vol. 44(4) Oct-Dec. 1989 with kind permission of the publisher.

NOTE

1. V.K.R.V. Rao (1983) was the first to draw our attention to the rising input-output ratio in the case of agriculture and animal husbandry. See also Nadkarni (1988).

REFERENCES

Bhalla, G.S. and G.K. Chadha (1983). *Green Revolution and the Small Peasant,* Concept Publishing Company, New Delhi.

Bhalla, G.S. and D.S. Tyagi (1989). Patterns in Agricultural Development : A District Level Study, Institute for Studies in Industrial Development, New Delhi.

Bhalla, Sheila (1987). "Trends in Employment in Indian Agriculture, Land and Asset Distribution", *Indian Journal of Agricultural Economics,* Vol. 42. No. 4 October-December.

David, C.C., K. Otsuka and Yujiro Hayami (1989). "Priorities for Rice Research: A View from the DIS Project", Discussion Paper, International Rice Research Institute, Manila, Philippines.

Desai, D.K. (1988). "Institutional Credit Requirements for Agricultural Production-2000 A.D.", *Indian Journal of Agricultural Economics,* Vol. 43, No. 3 July-September.

Government of India, (1989). National Accounts Statistics 1980-81 to 1986-87 (New Series), Central Statistical Organisation, Department of Statistics, Ministry of Planning, New Delhi.

Grabowski, Richard (1979). "The Implications of Induced Innovation Model", *Economics Development and Cultural Change,* Vol. 27, No. 4, July.

Grabowski, Richard (1981). "Induced Innovation, Green Revolution, and Income Distribution: Reply", *Economic Development and Cultural Change,* Vol. 30, No. 1, October.

Haque, T. and Sunita Verma (1988). "Regional and Class Disparities in the Flow of Agricultural Credit in India", *Indian Journal of Agricultural Economics,* Vol. 43 No. 3, July-September.

Hayami, Yujiro (1981). "Induced Innovation, Green Revolution, and Income Distribution : Comment", *Economic Development and Cultural Change,* Vol. 30, No. 1, October.

Hayami, Yujiro and Vernon W. Ruttan (1985). *Agricultural Development : An International Perspective,* Revised Edition, The Johns Hopkins University Press, Baltimore, U.S.A.

Indian Council of Agricultural Research and International Rice Research Institute (ICAR and IRRI) (1977). Report of Production Oriented Survey on Rice in Problem Areas of Bihar, Orissa and West Bengal, New Delhi and Manila, Philippines, September.

International Rice Research Institute (IRRI) (1982). A Plan for IRRI's Third Decade, Manila, Philippines.

Janvry, Alain de and Jean-Jacques Dethier (1985). Technological Innovation in Agriculture : The Political Economy of its Rates and Bias, Consultant Group on International Agricultural Research (CGIAR) Study Paper No. 1, The World Bank, Washington, D.C., U.S.A.

Jose, A.V. (1988). "Agricultural Wages in India", *Economic and Political Weekly*. Vol. 23 No. 26, June 25.

Jodha, N.S. (1989). "Dry Farming Research : Issues and Approaches", in N.S. Jodha (Ed.) (1989). *Technology Options and Economic Policy for Dryland Agriculture : Potential and Challenge*, The Indian Society of Agricultural Economics, Bombay : Concept Publishing Company, New Delhi.

Mellor, John W. (1988). "The Intertwining of Environmental Problems and Poverty", *Environment*, November.

Nadkarni, M.V. (1988). "Crisis of Increasing Costs in Agriculture : Is There a Way Out?, *Economics and Political Weekly*, Vol. 23, No. 39, September 24.

Nayyar, Rohini (1987). "Female Participation Rates in Rural India", *Economics and Political Weekly*, Vol. 22, No. 51, December 19.

Pinstrup-Andersen, Per (1982). Agricultural Research and Technology in Economic Development, International Food Policy Research Institute, Washington, D.C., U.S.A., Longman, Harlow, Essex, U.K.

Prasad, C., B.N., Choudhary and B.B., Nayar (1987). First-Line Transfer of Technology Projects, Indian Council of Agricultural Research, New Delhi, September.

Raman, K.V.,T. Balaguru and R.K. Samanta (1989). "Agricultural Research Systems for the Future: The Indian Context", Papers presented at CASAS-NAARM Asia Regional Seminar, National Academy of Agricultural Research Management, Hyderabad (A.P.).

Rao, C.H. Hanumantha (1975). "Technological Change and the Distribution of Gains in Indian Agriculture". The Macmillan Company of India Ltd., Delhi.

Rao, C.H. Hanumantha (1976). "Factor Endowments, Technology and Farm Employment : Comparison of East Uttar Pradesh with West Uttar Pradesh and Punjab", *Economic and Political Weekly*, Vol. 11, No. 39. September 25.

Rao, C.H. Hanumantha (forthcoming). "Some Interrelationships between Agricultural Technology, Livestock Economy, Rural Poverty and Environment: An Inter-State Analysis for India", in Golden Jubilee Volume: *Agricultural Development Policy : Adjustments and Reorientation*, Indian Society of Agricultural Economics, Bombay; Oxford & IBH Publishing Co. Pvt. Ltd. New Delhi.

Rao, V.K.R.V., (1983). *India's National Income, 1950-1980: An Analysis of Economic Growth and Change*, Sage Publications India Pvt. Ltd., New Delhi

Research and Information System for the Non-Aligned and Other Developing Countries (RIS) (Ed.) (1988). Biotechnology Revolution and the Third World: Challenges and Policy Options, New Delhi.

Shergill, H.S. (1987). "Impact of New Technology on Farm Size-Productivity Relationship in Punjab Agriculture: A Decomposition Analysis", Department of Economics, Punjab University, Chandigarh (mimeo.).

Swaminathan, M.S. (1989). "Restructuring National Research Systems to Meet Agricultural Development Challenges", Papers presented at CASAS-NAARM Asia Regional Seminar, National Academy of Agricultural Research Management, Hyderabad (A.P.).

Tyagi, D.S. (1989). "Managing India's Food Economy: Alternatives in a Self-Sufficiency Environment", New Delhi (mimeo.).

Vaidynathan, A. (1986). "Labour Use in Rural India: A Study of Spatial and Temporal Variations", *Economics and Political Weekly*, Vol. 21, No. 52, December 27.

2

TECHNICAL CHANGE AND WHEAT PRODUCTIVITY IN POST-GREEN REVOLUTION PUNJAB

D.S. SIDHU & DEREK BYERLEE

Punjab is one of the most important wheat producing States in India. It covers 14 per cent of national wheat area, accounts for 25 per cent of national wheat production, and provides more than 50 per cent of the wheat procured by the government food distribution scheme. It also has the highest wheat yield in the country, averaging 3.6 t/ha in 1986-88 compared to 2 t/ha for India as a whole.

Punjab was the first State to widely adopt the new wheat technology associated with the Green Revolution of the late 1960s. By the mid-1970s, the process of adoption of new high-yielding wheat varieties (HYVs) was essentially complete. During this decade (1966-75), wheat production increased at the extraordinarily high rate of 10 per cent annually, with over half this increase contributed by yield gains (Fig. 2.1). In the following decade, the rate of production increase declined to 4.8 per cent per year, with yield gains slowing to 2.7 per cent per year.

To maintain rapid increases in productivity, especially for wheat, is a matter of great concern to researchers and policy-makers alike. The lower rate of increase in wheat yields in recent years raises a number of questions. First, what have been the major sources of growth in wheat productivity in the post-green revolution period after HYVs were widely adopted? Second what are the prospects for continuing to exploit these sources of growth? Third, to what extent does the slower growth in productivity reflect changing input-output price ratios that provide a disincentive for investing in improved technology? Fourth, how are

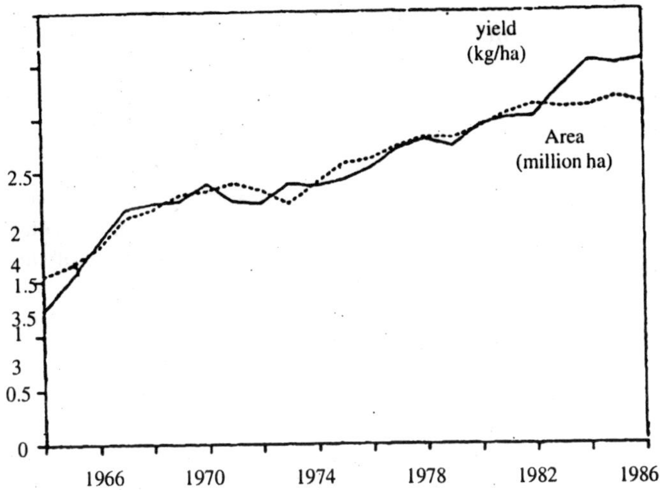

Fig. 2.1 : Trends in Wheat Area and Yield

increased yields and changing practices reflected in changes in costs of production and total factor productivity in wheat? And finally, to what extent has increased productivity in wheat been captured by producers or passed on to consumers in the form of lower prices?

Finding answers to these questions was the major motivation for this study, which provides a descriptive analysis of major changes in the Punjab wheat economy since 1971-72, after the widespread adoption of HYVs. The paper is based upon an analysis of trends in micro-level farm data collected under the project—Comprehensive Scheme for the Study of Costs of Cultivation of Principal Crops in the Punjab. This data set enables us not only to trace changes in input use and prices but also to compute an index of total factor productivity that considers changes in output in relation to changes in *all* inputs. This index is superior to the usual approach in India of equating productivity with yield per unit of land area, which does not account for the increasing importance of purchased inputs in substituting for land in wheat production (Fig. 2.2). In addition, the total factor productivity index has been proposed as way of measuring sustainability—an issue of concern in the intensive production systems of the Punjab (Chopra 1990; Byerlee and Siddiq 1990).

Data Sources and Methods

The study of Costs of Cultivation has run continuously since the 1960s. However, prior to 1972 the sample size was often small or restricted to certain districts. Because of this limitation and our interest in the post-green revolution period, only data since 1971-72 are analysed in this study.[1] In these surveys, enumerators residing in the villages keep daily records of farm operations under the close scrutiny of supervisory staff. The sampling method and approach have varied over the years (see Appendix A), but the surveys have generally included 200 farmers for all the Punjab up to 1985-86, and 300 farmers thereafter. The variation in sampling approach and coverage implies that some caution should be used in interpreting time trends in variables. However, the correlation between surveyed yields and official yield statistics was 0.92. Similarly, in the case of fertilizer the correlation between survey estimates and statistics on fertilizer off-take in the rabi (winter) cycle is also very high ($r = 0.95$).[2]

The survey data provide information on yields, human labour, machine costs, animal labour, fertilizer use, irrigation charges, pesticide use (mainly herbicides), and land rental charges (for full details see Appendix A). Price variables were deflated by the wholesale price index to a 1987 base. Trends in variables were computed through log-linear time trend regressions.

Finally, a Total Factor Productivity Index (TFPI) was computed. It is defined as the ratio of the output index to the input index. The input index was constructed as an index of all input quantities valued by their prices in a base year. That is

$$\text{Index (input)} \sum_i p_{i,o} X_{i,t} / \sum_i p_{i,o} X_{i,o}$$

where $p_{i,o}$ is the price of input i in period $t = 0$ and X_{it} is the physical quantity of input i used in period t. The output index was estimated simply as the yield in period t divided by the yield in the base period. The TFPI was then calculated as the ratio of the output index to the input index.

As with all methods using index numbers, the results may be sensitive to the choice of base year. In this case we computed the input index using both 1972 and 1987 prices as the base but the differences were very small; hence only the index based on 1987 prices is reported.

Changes in the TFPI are a measure of the effects of technological change. Since purchased inputs, such as fertilizers make up an ever increasing share of the total cost of wheat production, it is superior to the usual method of computing productivity as yield per unit of land.

Normally the TFPI is calculated for all farm enterprises combined, although some authors have argued for more disaggregation by type of enterprise (Huffman 1988). In our case, we have computed the TFPI for the major crop, wheat, which occupies over 80 per cent of the area sown in the 'rabi' cycle. This may over- or underestimate TFPI for all farm enterprises to the extent that there are differential rates of technical change across the major farm enterprises or changes in TFPI for wheat have negative or positive effects on other crops in the system (through delayed planting of the subsequent crop in the rotation, e.g., delayed sowing of wheat due to potato-wheat, paddy-wheat rotations). Also, the TFPI for what will under- estimate the TFPI for all enterprises if cropping intensity has increased significantly.

Tends in Inputs and Outputs

Overall trends in inputs and yields are summarised in Table 2.1. Inputs can be broadly classified as follows:

TABLE 2.1 : Overall trends in inputs and outputs

	Land Saving	Labour Saving
Traditional	Seed quantity Organic manure	Bullocks
Modern	HYV Fertiliser Pesticide	Machine labour (tractorisation)

Pesticides may also be classified as labour saving inputs if their primary role is to substitute for labour without an associated increase in yields (e.g., use of herbicides instead of hand weeding).

It is clear from Table 2.1 and Fig. 2.2 that modern inputs have substituted for traditional inputs. In the period under analysis, Punjab farmers completed the switch from local varieties to HYVs. In 1972, 73 per cent of farmers used HYVs and by 1976 this percentage had increased to 90 per cent (Fig. 2.2). The transition to HYVs was essentially completed by 1980 although farmers have adopted several newer higher yielding varieties, such as WL711 and HD2309 since the

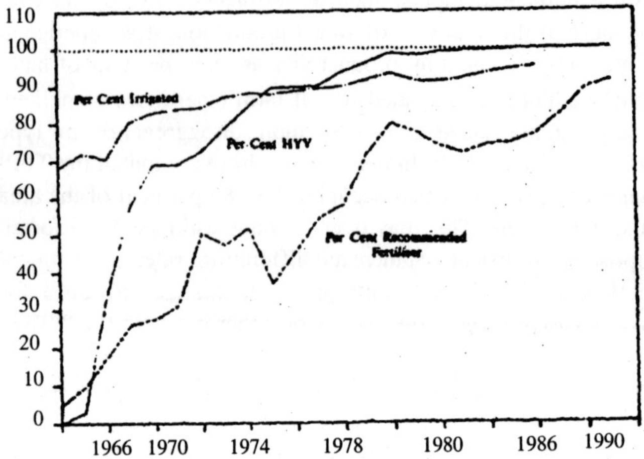

Fig. 2.2 : Trends in Input Use in Wheat

original HYVs were introduced. Also, the wheat area under irrigation increased from 84 per cent in 1972 to over 95 per cent at the end of the period. Finally, by 1987 average fertiliser use reached over 90 per cent of the recommended dosage.

In contrast to the use of HYVs and growth in irrigated area, which had reached almost saturation levels early in the period, fertiliser use continued to increase rapidly, from about 100 kg nut/ha in 1972 to 180 kg nut/ha in 1987—a growth rate of 4.4 per cent per year. The general fertiliser recommendation for wheat in the Punjab is 125-62-30 kg/ha of N-P-K, or a total of 217 kg/ha. Advanced districts, such as Ludhiana, have already exceeded this recommendation, at least for nitrogen. In addition, the marginal productivity of additional fertiliser use has also slowed and was estimated to be 5.6 kg wheat per kilogram of nutrient (Grewal and Rangi 1983). Hence, growth in fertiliser use and its marginal contribution to yield increases is expected to be substantially lower in the future than during the period under study.

In the 1980s, the use of pesticides, largely herbicides, expanded rapidly. The widespread losses caused by the weed Phalaris minor have been a major impetus for increased herbicide use. In Ludhiana district it is estimated that over two-thirds of the wheat area is now treated with herbicide. Since Phalaris is difficult or impossible to control manually and has caused serious yield losses, it is reasonable to assume that using herbicides to control Phalaris has been a land saving rather than a labour saving innovation.

Of the traditional inputs only seed rate has shown a steady increase from about 80 kg/ha in 1972 to 110 kg/ha in 1987. A marginally higher seed rate (which surpasses the recommended rate of 100 kg/ha) has been observed in other post-green revolution settings (Byerlee and Siddiq 1990; Traxler 1990). Higher seed rates probably reflect the complementarity between good plant stand and inputs such as fertiliser and water and also provide insurance against poor germination.

In contrast there is evidence that use of organic manure has declined over time (Table 2.2), a finding that is in accord with other reports (Chopra 1990). This decline has occurred in part because cropped area has expanded faster than livestock numbers, since bullocks have been replaced by tractors. As the human population has grown, a higher proportion of organic manure may now also be used for fuel. The use of chemical fertilizers has partly substituted for the application of organic manure, by providing nutrients, but the reduction in use of organic manure may have negatively affected soil physical properties such as the rate of water infiltration.

TABLE 2.2 : Trends in Input use and Wheat Yields, 1972-87, Punjab

	Period				*Trend 1972-87*
	1972-75	*1976-79*	*1980-83*	*1984-87*	
Traditional inputs					
Seed (kg/ha)	81	92	96	106	2.16*
Manure (t/ha)	1.41	1.19	0.74	0.92	–4.11
Modern inputs					
Fertiliser (kg. nut/ha)	102	125	164	168	4.36*
Pesticide (Rs. 1987/ha)	11	5	35	92	27.61*
Labour and machinery					
Human labour (h/ha)	522	457	393	431	–1.71*
Animal labour (h/ha)	101	76	37	26	–12.24*
Human labour less animal labour (h/ha)	421	381	356	405	–0.30
Machine labour (Rs. 1987/ha)	407	484	643	679	4.76*
Wheat yield (100 kg./ha)	25.2	24.0	28	32.3	2.32*

* denotes signifacance at the 1 per cent level.

Turning now to the labour saving inputs, it is clear that machinery labour has rapidly substituted for human and animal labour over the period under analysis. The most dramatic change has occurred in animal labour, which declined by about 80 per cent over the period after tractors largely replaced the use of bullocks in preparing land, planting, and threshing. The number of tractors in the Punjab increased from only 10,000 in the mid-1960s to 2,30,000 in the late 1980s. In addition, electric and diesel tubewells have completely replaced animal power in pumping water throughout the Punjab.

Although the Green Revolution was widely associated with an increase in employment, the labour input per hectare has actually fallen by 1.7 per cent annually in the post-green revolution period (Table 2.2). This decline appears to be largely associated with the decline in animal power. Assuming that one tractor hour is equivalent to eight animal hours in land preparation, the calculated trend in labour use net of animal power is non-significant. Human labour is now mainly used for irrigation and harvesting. Irrigation labour is independent of yields and, although labour required for harvesting is expected to have increased with yields, the spread of threshers, and more recently combine harvesters, may have negated this effect.[3]

Tends in Input and, Output Prices

The real prices of most inputs declined from 1972 to 1987. In the case of fertiliser this partly reflects the high prices prevailing in 1975 and 1976. Excluding these years, however, the ratio of fertiliser prices to wheat prices has been remarkably constant over the period (Fig. 2.3).

Although real wage rates increased significantly in the Punjab during the Green Revolution period of the late 1960s (Jose 1988). This trend has been reversed since then. However, the decline in real wages (Table 2.3) is small and not statistically significant. Stagnating real wages are constant with the fall in labour input per hectare as a result of mechanization, coupled with net immigration from other states to rural Punjab.

Likewise the real cost of animal power has not changed over the period. However, the price of machinery services has fallen and together with increased cropping intensity has provided the impetus to substitute machinery services for animal power.

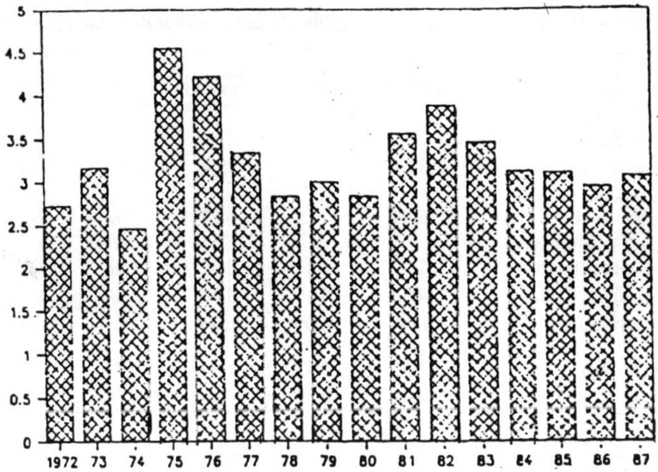

Fig. 2.3 : Ratio of Fertiliser Price to Wheat Price

Finally, the lack of a significant trend in land rental values is interesting (Table 2.3). It is often assumed that the surplus generated by technical change will be capitalised in increased prices for production factors whose supply is inelastic, especially land. But in the case of the Indian Punjab there is no evidence of this effect (Kahlon and Kurien 1984). However, real land rental values do seem to have increased significantly in the Punjab of Pakistan (Renckow 1990).

Against these changes in input prices, real wheat prices have fallen consistently every year except 1974, at an average rate of 3.5 per cent annually. A declining real output price is consistent with rapid technical change in a closed economy and suggests that the benefits of technical change have been captured more by consumers than producers, a theme which is explored more fully below.

Note, however, that there has been no trend in the real price of straw (Table 2.2), so that the grain-to-straw price ratio has decreased over time. In contrast, in the Punjab of Pakistan real straw prices declined over this period as straw production increased faster than animal numbers (Byerlee and Iqbal 1987).

TABLE 2.3 : Trends in Real Input use and Output Prices in Wheat Production, 1972-87

	Period				*Trend*
	1972-75	*1976-79*	*1980-83*	*1984-87*	*1972-87*
Traditional inputs					
Seed (Rs./kg)	2.92	2.95	2.54	2.17	–2.46*
Manure (Rs./100 kg)	31.4	27.2	19.4	18.3	–4.89*
Modern inputs					
Fertiliser (Rs./kg)	8.44	7.75	6.65	5.32	–3.48*
Labour and machinery					
Human wage (Rs./h)	2.04	2.21	2.02	1.98	–0.43
Animal wage (Rs./h)	4.99	6.14	5.58	5.84	0.83
Land rental (Rs./ha)	1747	1599	1517	1687	–0.48
Wheat					
Grain (Rs./kg)	2.65	2.32	1.94	1.74	–3.47*
Straw (Rs./kg)	0.22	0.26	0.20	0.24	–0.20

* denotes signifacance at the 1 per cent level.

Trends in Cost Structure

Given the input use and price trends noted earlier, one would expect to see significant shifts in the cost structure for wheat production. In particular, the share of animal labour in the total cost of wheat production decreased sharply in the period 1972-87 from 9.1 to 2.7 per cent and the share of labour decreased from 19.1 to 15.6 per cent while the share of machinery labour increased from 12.6 to 22.8 per cent (Table 2.4). Likewise the share of chemical inputs also increased over the period.

TABLE 2.4 : Trends in Cost Structure for Wheat, 1972-87

	Period			
	1972-75	*1976-79*	*1980-83*	*1984-87*
Per cent of total costs due to				
Seed and manure	5.0	5.3	4.8	4.5
Fertiliser and pesticides	15.2	16.6	21.1	18.0
Human labour	19.1	17.5	14.9	15.6
Animal labour	9.1	8.0	3.9	2.7
Machine labour[a]	12.6	17.8	22.2	22.8
Irrigation charges	5.9	5.0	2.7	3.8
Land rent	31.2	27.7	28.4	30.7
Total	100	100	100	100
Per cent cash costs	50.1	55.0	62.9	61.2
Per cent non-cash costs[b]	49.9	45.0	37.1	38.8
Total	100	100	100	100

a. Includes depreciation and interest.

b. Includes seed, manure, family labour, animal labour, own machines, and own land.

The substitution of chemical inputs and machinery for labour and land also increased the share of cash costs in total production costs from 50 per cent in 1972-75 to 61 per cent in 1984-87. Cash costs rose most rapidly in the 1970s with the rapid adoption of machinery services and increased fertilizers inputs. Even for the traditional inputs of land and labour there has been a tendency towards increased cash expenditures, with hired labour increasing its share relative to family labour and a tendency for a higher share of land to be rented. Hired labour accounted for 50 per cent of total labour in 1971-72 and 59 per cent in 1986-87. The leased-in land on rental basis was 11 per cent in 1971-72 and increased to 19 per cent in 1986-87.

Costs, Prices and Margins

As expected with rapid technical change, the cost of production in real terms decreased steadily at a rate of 2.9 per cent annually (Tables 2.5 and 2.6, Fig. 2.4 and Appendix B). (The cost of production used here is what is known as 'measure C', which is calculated with all costs, including the imputed value of family labour and land) Although some have questioned whether the cost of wheat production has indeed declined (Kahlon 1984), more recent data suggest the decline observed here is consistent with observations for wheat in other states of India but not generally for other crops (Mruthynjaya and Kumar 1989). Nonetheless, a declining real cost of production of 2.9 per cent annually against a decrease in real wheat prices of 3.6 per cent annually implies some squeezing of the profit margin to farmers. Hence the real margin per 100 kg of wheat produced (in 1987 rupees) has fallen from Rs. 60/100 kg in 1972-75 to only Rs. 22/100 kg in 1984-87 (Table 2.5). Despite a

TABLE 2.5 : Costs and Returns in Wheat Cultivation, 1968-87

	Period				
	1968-71	*1972-75*	*1976-79*	*1980-83*	*1984-87*
Gross income (Rs./ha)	2073	2586	2955	4207	5859
Total cost (Rs./ha)	1518	2034	2672	3696	5055
Profit margin (Rs./ha)	555	552	283	511	834
Per cent profit margin[a]	38	28	10	15	15
Cost/100 kg (Rs./100 kg)	60	72	103	118	138
Procurement price (Rs./100 kg)	76	94	113	135	159
Profit margin (Rs./100 kg)	16	22	10	17	21
Real margin/100 kg (Rs./100 kg)	61	60	21	24	22
Real margin/ha (Rs./100 kg)[b]	1724	1522	509	714	739

a. Per cent profit marign over cost.

b. Constant 1987 Rs.

Source : Sidhu (1979).

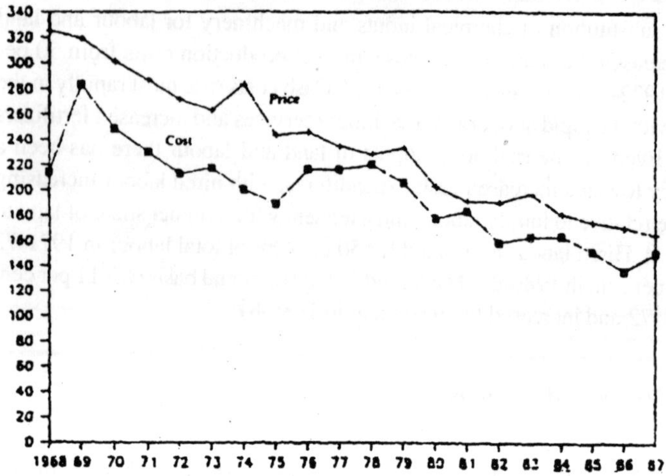

Fig. 2.4 : Procurement Price and Cost of Production for Wheat Production

significant increase in yield, the real margin/hectare has also fallen by half over this period.[4] Likewise, the percentage margin of the sales price of wheat over the production cost fell from 36 per cent in 1968-71 to less than 20 per cent for the recent decade, 1976-87 (Fig. 2.5).

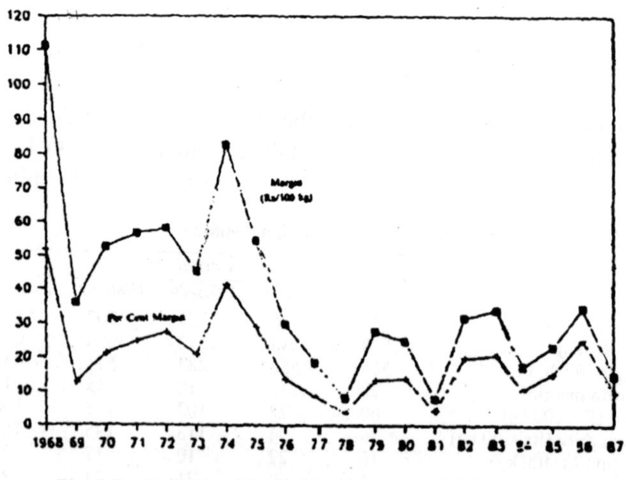

Fig. 2.5 : Real Price Margin (Rs./100 kg) and Per cent Margin Over Costs in Wheat Production

These changes are consistent with the hypothesis that farmers initially gained most of the benefits of using the new wheat technology, especially between 1968-1975. However, since 1975 government policy appears to have favoured consumers, with the procurement price being set to exploit the downward trend in production costs brought about by technical change. The lack of increase in land rental values also supports the conclusion that over the long run farmers have not been able to capture the benefits of technical change.

Another observation that can be made from the data is that although producer prices have been quite stable (with an average coefficient of variation around trend of 3.8 per cent) and average yields have also shown low variability (CV = 9 per cent), the returns to wheat production have shown a high variability of 55 per cent around trend (Table 2.6).[5] Much of this variability occurred during the early 1970s when fertiliser, fuel and grain prices were particularly volatile. Since 1976 the CV of net returns in wheat production has still averaged a high 44 per cent. Hence, stabilization of producer prices does not appear to have been very effective in reducing variability in farmers' incomes.

TABLE 2.6 : Trends in Procurement Price, Cost of Production and Margins in Wheat Production, 1972-87

Trend in Real Price/Cost	*Per cent/Year*	*Coefficient of Variation*
Total cost/100 kg	–2.93*	7
Procurement price	–3.58*	5
Profit margin/100 kg	–6.64**	55
Profit margin/ha	–4.32	43
Percentage margin	–3.72	56

Notes : **, * denote significance at the 5 per cent and 1 per cent levels, respectively

Factor Productivity

Fig. 2.6 and Table 2.7 provide partial productivity measures—that is, wheat production per unit of a given input. During the Green Revolution period the use of the new seed-fertiliser technology tended to increase land productivity—that is, it was a labour intensive, land saving technology (Hayami and Ruttan 1985; Sidhu 1974). In the more recent period under study, land productivity has continued to increase at a slower rate of 2.3 per cent per year but the largest gain has been in labour productivity which has increased by 4 per cent per year, reflecting the dramatic increase in mechanisation. At the same time productivity

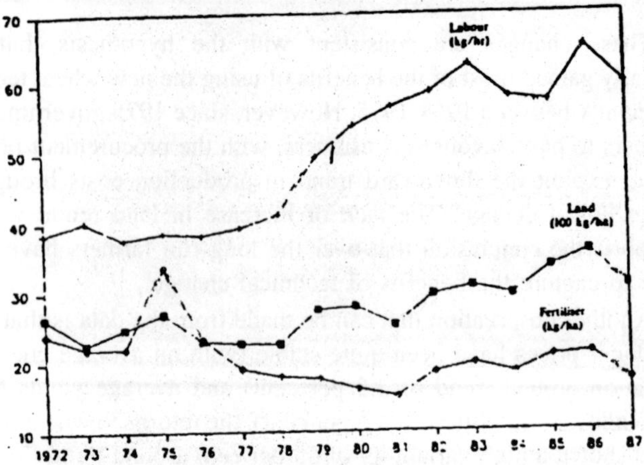

Fig. 2.6 : Partial Productivity Measures for Land Labour and Fertiliser

per unit of fertiliser applied has tended to decline as the marginal returns to additional fertiliser use have fallen. (Grewal and Rangi 1983).

The estimated total factor productivity index (TFPI) is also given in Table 2.7 and Fig. 2.7. The interesting result of these calculations is that the input index has hardly changed over the period and shows no significant trend. The increase in chemical and machinery inputs has been more than

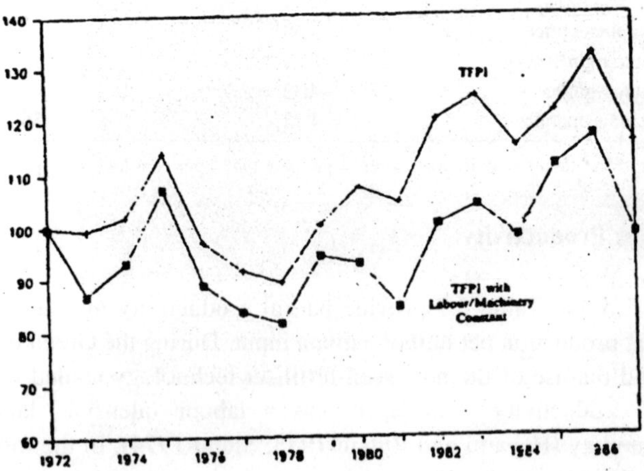

Fig. 2.7 : Index of Total Factor Productivity (TFPI) in Wheat Production

TABLE 2.7 : Trends in Indices of Partial Productivity and Total Factor Productiviy

	Period				*Trend*
	1972-75	*1976-79*	*1980-83*	*1984-87*	*1972-87*
Partial productivity					
Land (100 kg/h)	25	24	28	32	2.32***
Labour (kg/ha)	39	42	58	60	4.03***
Fertiliser (kg/kg nut)	25	20	.17	19	–2.03**
Total factor productiviy (TFPI)					
Input index	92	96	93	100	0.57**
Output index	95	91	107	122	2.32***
TFP index	104	95	115	122	1.75***
Adjusted TFPI	102	89	102	110	1.05*

*, **, *** denotes significance at the 10 per cent, 5 per cent and 1 per cent level, respectively.

cancelled by a decline in labour and animal power inputs.

Because the input index has changed little, the total factor productivity index is highly correlated with yield. Overall the TFPI has risen by 1.7 per cent per year which is consistent with long-term changes in total factor productivity of 1-2 per cent annually in the US (Ball 1985; Huffman 1988) and overall estimates for India [Evenson and McKinsey (nd)] and for Pakistan (Wizarat 1981).

Nonetheless, about half of the increase in productivity has been due to increased labour productivity as a result of mechanisation. This can be seen by assuming that mechanisation has not led to any increase in yields, an assumption consistent with the evidence from Agarwal (1984) and Binswanger (1978) and by recalculating the TFPI using constant animal and machinery inputs of 1972. The result is a much slower increase in productivity of 1 per cent per year. It is sobering that practically all this increase occurred during the latter part of the period 1983-87. That is, the land augmenting components of technical change—new varieties, increased fertiliser use and adoption of pesticides—have had relatively little impact of TFP for much of the period. It is still too early to determine the factors which have reversed stagnating productivity in recent years without a dramatic change in input use of new HYVs, but this reversal may relate to an increase in input efficiency as farmers have learned how to manage the new technology better (Byerlee 1987).

Conclusion

This review of technical change and wheat productivity in the 1970s and 1980s reveals a number of interesting trends. First, the Green

Revolution has been consolidated through further intensification in the use of modern land-saving inputs, especially fertiliser and herbicides. Use of those inputs continued to increase strongly even after the full adoption of high yielding varieties. However, the use of labour saving technologies, especially tractors, has expanded even more rapidly in this period. Labour saving technologies have reversed the trend toward higher labour inputs in wheat production that was observed when the new wheat technology was first introduced.

No major changes in relative input prices were observed over the period. However, the trend toward high real wages that characterised the Green Revolution period is no longer evident. The reduced labour input in wheat production may be one reason for this change.

A striking trend is the consistent fall in the real price of wheat received by producers. Although the falling cost of production induced by technical change has partly compensated for the decline in real wheat prices the real margin to wheat producers has also fallen sharply. There is evidence that during the Green Revolution period farmers were the beneficiaries of much of the surplus generated by increased wheat productivity, but in the 1970s and 1980s wheat consumers have received the bulk of the benefits.

The calculated changes in total factor productivity support these findings. The most important source of productivity increases has been labour saving technology. The use of biochemical technologies has only modestly increased total productivity and that occurred mostly in the latter part of the study period, although it is too early to conclusively identify the specific source of this productivity gain.[6] We have not analysed changes in the resource base (except for an observed decrease in use of organic manures), but there seems little reason at this stage to be concerned about the sustainability of productivity levels that have been achieved in wheat production. However, given the high level of inputs used and the narrowing gap between the highest and lowest wheat yields obtained by farmers (Singh *et al,* 1987) and between farmers yields and yields on the research station, future sources of yield gains are not readily evident, at least at the rate observed in the recent past. Undoubtedly, further productivity gains in wheat production will have to be achieved by more efficient use of inputs such as fertiliser and water.

APPENDIX –A

SAMPLING METHOD

To analyse wheat productivity, farm-level production data from 1971-72 to 1986-87 for 300 randomly selected farmers[7] were collected from the project. The data were collected through the cost accounting method; full-time field investigators residing in the villages kept daily records of farm operations.

Prior to 1985 the sample was selected by three-stage stratified random sampling with a tehsil (sub-district) as the primary sampling unit, a cluster of three villages as the second stage unit and an operational holding within the cluster as the third stage. From 1985 the state of Punjab was demarcated into three homogenous zones based on cropping pattern, irrigation facilities, rain-fall, and soil type. The first zone is the rice-what area, located in the northern part of the state. It includes 22 tehsils and covers 41.8 per cent of Punjab's cultivated area. The second zone is the maize-wheat area, which comprises the central part of the state and includes 13 tehsils and covers 27.6 per cent of the cultivated area. The third zone is the cotton-wheat area, located in the south-western part of the state. It includes 10 tehsils and covers 30.5 per cent of cultivated area.

Thirdly tehsils were selected from these three zones on the basis of the area planted to four crops (one of which was wheat). One village from each of those tehsils was selected with probability proportional to area under the four study crops. Then a cluster of three villages was formed by selecting one contiguous village to the south and one to the west of the nucleus village. All operational holdings of the villages in each cluster were enumerated and classified into five farm size classes. From each size class two cultivators were selected randomly to create a sample of 10 cultivators from each cluster and a total sample of 300 holdings for this study.

Prior to 1985-86 a rotational sample of 200 holdings was randomly selected throughout the state. During 1971-72 wheat was the principal crop under study. Only a sub-sample of wheat producers was used in the 1972-73 and 1973-74 surveys. However, data on wheat continued to be collected from cotton and sugarcane farmers to form 200 observations each year.

Variables for which Data were Collected

Wheat Yield: The total production of wheat in the sample was divided by area under wheat to give the yield per hectare.

Human Labour: This input was measured in adult units of hours spent in wheat production operations. A full account of day-to-day field operations done by family or casual and permanent hired labour was kept for each farm. Casual and permanent hired labour were valued at the actual wages paid and the value of family labour was imputed at the rates paid to permanent labour.

Bullock Labour: Bullock labour was also measured in hours of work in each operation in wheat production. Bullock labour was valued at the market rate.

Machine Labour: Since different types of machines with different horse power were used, machine labour was recorded in value units to obtain a uniform unit of measurement over time. Machine inputs were estimated by deflating the value of machine expenditures by the wholesale price index for tractors.

Fertiliser: The quantities of fertilisers in nutrient form were totalled to form one variable. Potassium accounted for only a small proportion of the total.

Irrigation: Irrigation was provided by tubewells and canals. To put this variable on a measurable basis, it was decided to specify it in the value terms. For farmers irrigating their crop with tubewells, the variable costs and the fixed cost of irrigation equipment in terms of depreciation and interest were taken into account. The expenditure on irrigation was apportioned according to the number of hours spent irrigating each crop. For canal-irrigated farms the water charges for wheat were taken. However, tubewells run by electricity were the main source of irrigation and farmers were charged a flat rate that does not reflect true irrigation costs.

Pesticides: Farmers used several chemicals on wheat but herbicides predominated. The value of each chemical applied by the farmer was summed to arrive at the total expenditure on chemicals. Changes in input use were estimated by deflating by the price index for agricultural chemicals.

Land: Owned land was valued through imputing land rent on the basis of leased-in land. Land rent was, however, subjected to a maximum limit of one-third of the total value of the main crop and by-products. Thus a rent known as 'fair rent' was entered in the records rather than the true market rent. However, in most cases the difference between the two rent figures is believed to be small.

APPENDIX –B

COST OF PRODUCTION, PROCUREMENT PRICE AND PROFIT MARGIN FOR WHEAT, 1968-87

Total Cost (Rs./100 kg.)	*Procurement Price (Rs./100 kg.)*	*Profit Margin (Rs./100 kg.)*	*Per cent Margin*	*Real Cost[a] (Rs./100 kg.)*	*Real Margin[a] (Rs./100 kg.)*
50.0	76	26.0	51.9	214.1	111.2
67.5	76	8.5	12.7	283.3	35.9
62.7	76	13.3	21.2	248.2	52.7
61.0	76	15.0	24.7	229.9	56.7
59.7	76	16.3	27.3	213.2	58.2
67.1	81	13.9	20.7	217.8	45.1
74.3	105	30.7	41.2	200.7	82.8
87.8	113	25.2	28.8	189.2	54.4
99.5	113	13.5	13.6	216.8	29.5
101.4	110	8.6	8.5	216.5	18.4
108.6	113	3.9	3.6	220.4	8.0
101.5	115	13.5	13.4	205.9	27.5
102.8	117	14.2	13.9	178.1	24.7
124.7	130	5.3	4.3	183.5	7.8
118.5	142	23.5	19.8	158.9	31.5
125.2	151	25.8	20.6	163.5	33.7
137.5	152	14.5	10.6	164.1	17.3
136.3	157	20.7	15.2	151.9	23.0
129.3	162	32.7	25.3	136.3	34.5
150.4	165	14.6	9.7	150.4	14.6

Note : a deflated by wholesale price index to 1987 Rs.

Reproduced from *Economic and Political Weekly*, December 28, 1999 with kind permission of the publisher.

NOTES

1. For an analysis of earlier years, see Sidhu (1979).
2. The differences between survey yields and official statistics were most pronounced in the 1970s. In the period 1972-75, survey yields were 9 per cent above official yields and in 1976-78 there were 7 per cent under. Since 1978 this difference has not exceeded 2 per cent. This variation undoubtedly reflects the varying sampling methods used in the 1970s in the cost of production surveys; sampling was standardised in the last decade.
3. The additional area under wheat in period 1972-90 has been about 8,00,000 ha which is 25 per cent of the total area under wheat (3.2 millions ha); harvest combines had a capacity of harvesting about 10 per cent of total wheat area in 1989-90.
4. See also Singh *et al* 1987 and George *et al* 1983.
5. The CVs here are calculated around the linear time trend.

6. Also, the analysis has not taken into account the effect of marginal lands coming under wheat and the effect of delayed sowing of wheat due to new rotations that were followed over time. As such it is a partial analysis of wheat productivity and not of the farming system as a whole.
7. The number of farmer respondents was 200 from 1971-72 to 1980-81. After each three-year-period, the sample was completely changed. Since 1980-81 the number of farmer respondents has increased to 300 and the sample continued to be changed after three years.

REFERENCES

Agarwal, B. (1984), 'Tractors, Tubewells and Cropping intensity in the Indian Punjab', *Journal of Development Studies*, 20:290-302.

Binswanger, H. (1978), *The Economics of Tractors in South Asia, An Analytical Review*, Agricultural Development Council, International Centre for Research in the Semi-Arid Tropics, New York/Hyderabad.

Ball, V.E. (1985), 'Output, Input, and Productivity Measurement in US Agriculture, 1948-79'. *American Journal of Agricultural Economics,* 67 : 475-86.

Byerlee, D. and M., Iqbal (1987), 'Recent Trends and Variability in Prices of Selected Crop and Livestock Products in Pakistan', *Pakistan Journal of Agricultural Sciences,* 1:158-168.

Byerlee D. (1987), *Maintaining the Momentum in Post-Green Revolution Agriculture: A Micro-Level Perspective from Asia,* International Develop- ment Paper No. 10, East Lansing: Michigan State University.

Byerlee, D. and A., Siddiq (1990), *Sources of Growth in Wheat Yields in Pakistan's Punjab 1965-2000: Is There a Sustainability Issue*? CIMMYT Economics Working Paper No. 90/04, Mexico DF, CIMMYT.

Chopra, K. (1990), *Agricultural Development in the Punjab: Issues in Resource Use and Sustainability,* New Delhi.

Evenson, R. E. and J., McKinsey (nd), Research, Extension, Infrastructure and Changes in Indian Agriculture, Draft paper, New Haven, Yale University.

George, M., V. N. J. Kurian and C. C. Mohan (1983), 'Factor Shares in Indian Agriculture: Temporal and Spatial Variations and Their Implications' *Indian Journal of Agricultural Economics,* 38:399-407.

Grewal, S.S. and P.S. Rangi (1983), 'An Analytical Study of Growth of Punjab Agriculture', *Indian Journal of Agricultural Economics*, 38(4):509-520.

Hayami, Y. and V. W. Ruttan (1985), *Agricultural Development: An International Perspective,* Johns Hopkins University Press, Baltimore.

Huffman, W. (1988). 'State Regional, and Activity Specific Agricultural Productivity Indexes'; Paper presented to joint ERS/IR-6 Conference on Agricultural Productivity Measurement Procedures, March 31-April I. Washington DC.

Jose, A. V. (1988), 'Agricultural Wages in India', *Economic and Political Weekly* (June), A46-A58.

Kahlon, A.S. and J. Kurien (1984), 'Trends in the Share of Rental Value of Land in Cost of Cultivation of Major Crops in India', *Agricultural Situation in India* (4):3-7.

Kahlon, A.S. (1984), *Modernisation of Punjab Agriculture,* Allied, New Delhi.

Mruthyunjaya and P. Kumar (1987), 'Crop Economics and Cropping Pattern Changes', *Economic and Political Weekly* (December), A159-A166.

Renkow, M (1990), *Land Prices, Land Rents, and Technological Change: Evidence from Pakistan*, CIMMYT Economics Programme Working Paper, Mexico DF, CIMMYT (forthcoming).

Sidhu, S.S. (1974), 'Economics of Technical Change in Wheat Production in the Indian Punjab; *American Journal of Agricultural Economics*, 56:217-226.

Sidhu, D.S. (1979), *Price Policy for Wheat in India*, Chand, New Delhi.

Singh, I.P.B. Singh and H.S. Bal (1987), 'Indiscriminate Fertiliser Use Vis-a-Vis Groundwater Pollution in Central Punjab.' *Indian Journal of Agricultural Economics*, 42 : 404-409.

Traxler, G. (1990), 'The Economics of Crop Management Research', Ph. D. thesis, Iowa State University, Ames.

Wizarat, S. (1981), 'Technological Change in Pakistan's Agriculture: 1953-54 to 1978-79', *The Pakistan Development Review* 20 : 427-445.

3

PRODUCTIVITY GROWTH

Scale and Technology Analysis of Punjab Wheat Farms

RENU KALLIANPUR & ROMESH DIWAN

Introduction

The total world production of food is sufficient to satisfy the food needs of all the people in the world.[1] Still, there are a large number of undernourished people in the world[2] basically because such a large population is poor[3]. Furthermore, under the existing methods and policies of production and distribution of food, there is no prospect for the elimination of undernutrition in the foreseeable future.[4] Next to human survival and nuclear war, this is the most important problem in the world.

The intellectual issue in this problem is how to produce more food ensuring, at the same time, that it reaches those who need it. The conventional wisdom, developed over the past 30 years, maintains that increased production follows from new technological inputs. On the other hand, there is an equally strong evidence that small farmers are more productive. There is an ongoing debate among economists how to reconcile these ideas and facts.

This paper enjoins this debate and makes the following contributions:

(i) It offers, for the first time, a precise and quantifiable definition of biological technical change;
(ii) It develops a model that can be estimated from easily available and generally reliable data;

(iii) It uses, for the first time, a new set of most recent data; and
(iv) It formulates and tests a hypothesis that reconciles the seemingly contradictory evidence on the relationship between land productivity and farm size.

Technology and Structural Theories of Land Productivity

The production of food depends upon a number of factors. The most important is land. However, a major part of the easily available land has been already brought into cultivation.[5] Future increases in production have to come from improvements in land productivity. The question is; what determines land productivity? Two schools of thought can be identified; call them—(i) technological, and (ii) structural.

The technology school is best represented by Schultz.[6] The basic argument of this school is as follows. Farmers, even in poor countries, are rational.[7] They produce at the efficiency frontier. It is the efficiency frontier itself that needs to be pushed up. The increases in efficiency frontier follow from the introduction of new technological inputs: seeds of high yielding varieties, fertilizers, irrigation facilities via tube-wells or large scale irrigation projects, tractors, herbicides, pesticides, etc. Agronomy studies working on these technological inputs provide evidence about the effectiveness of these inputs in raising land productivity.[8] Given the rationality of the farmer and the productivity of new technological inputs, the next step in the argument is straight-forward. The farmer should be provided these inputs at economic prices; i.e., at prices that will make the use of these inputs economic. Hence land productivity and food production will increase. Since Schultz initial work, a number of studies have provided evidence on this thesis.[9] By now, this thesis has become the prevailing conventional wisdom.[10] This school has conceptualized the numerous technological inputs into biological and mechanical technology.[11]

The 'structural' school is not as well structured or articulated. It maintains that labour both in its qualitative and quantitative aspects, is the major determinant of land productivity. It is so because the application of labour makes it possible to (i) improve the quality of land by such activities as levelling, ditching, land preparing, tilling, etc., (ii) perform various tasks at appropriate times such as watering, seeding, weeding, harvesting, etc., and (iii) take more intensive care of crops, such as weeding, pruning, etc. In economic language, the availability of quality labour input promotes the efficiency of the agricultural operations and investment in land quality.[12] The quality of labour input is positively related to the voluntary involvement by a person in the agricultural production on process. This is contrasted with the quality of labour input through wage or other compulsions.[13] Various forms of socio-economic relations promote, or discourage, the application of different levels of quantitative and qualitative labour input. Socio-economic relations interact with, and follow from, the socio-economic institutions. Examples of these institutions are land-ownership, tenurial arrangements, wage-market, co-operatives, etc. Some institutions encourage the applications of high quality and quantity of labour, such as family production units. Others discourage it. The land lord system discouraged such applications and has been considered to be both inefficient and unproductive.[14] The socio-economic structure, made up of institutions and relations, provides the necessary incentives and opportunities to work hard, make investments, etc. The 'structural' school has conceptualized the large number of socio-economic factors into "farm size". "Farm size" refers to the availability of a certain quantity of land for cultivation with a sufficient level of certainty over a medium or long period. The smaller farm size refers to a socio-economic structure that promotes the application of high quality and quantity of labour input thereby improving land productivity.

Farm Size and Land Productivity: An Inverse Relationship

The arguments from the two schools imply different relationship between land-productivity and farm size. The technology school argues that land productivity increases by an increasing use of new technological inputs. Some would argue that the technology is scale neutral. However, larger farms have much greater opportunity and resources to use proportionately larger quantities of new technological inputs. Writing $Z_{0,1}$ for land productivity and *FS* for farm size, the arguments of the technology school may be formulated in the hypothesis:

$$\frac{dZ_{0,1}}{dFS} > 0$$

Interestingly, the structural school comes to the opposite conclusion. It argues that the nature of socio-economic relations are such that small farms attract higher quality and quantity of labour input. The quality of the labour input is higher because the worker in a small farm is more likely to identify himself or herself with the production unit.[15] Large farms, on the other hand, have to use some form of compulsion to attract labour input; wage-labour is also a form of compulsion.[16] The quality of labour input may therefore be lower. Accordingly, the hypothesis[17] based on the arguments of the structural school may be written as,

$$\frac{dZ_{0,1}}{dFS} < 0$$

In view of its obvious importance, this hypothesis has generated a a number of empirical studies. Some of the studies based on data from India are listed in Table 3.1 below.

TABLE 3.1 : Relationship between Farm Size and Land Productivity : Empirical Evidence

Data Years	*Author & Study Year*	$dZ_{0,1}/dFS$	*Area Studied*
Pre-HYV			
1954-59	Sen (1962)	< 0	All India
1955-56	Mazumdar (1965)	< 0	2 Villages in UP
1955-56	H. Rao (1975)	< 0	3 Districts in Maharashtra
1954-57	A. Rao (1967)	?	3 Villages in P H and UP.
1959-60	Rudra (1968)	= 0	7 Districts in P H and UP
1957-64	Saini (1971)	< 0	All India
1954-57	Bhardwaj (1974)	< 0	All India
1955-68	Ghose (1981)	< 0	Bengal and Punjab
Post HYV			
1962-70	Rao & Chotigeat (1981)	= 0	Kerala and Tamil Nadu
1955-60/1966-70	H. Rao (1975)	= 0	Punjab, UP and AP
1954-70	Bhattacharya and Saini (1972)	?	All India
1962-73	Chattopadhyay and Rudra (1976)	may not hold	All India
1968-71	Bhalla (1979)	< 0	All India
1968-71	Deolalikar (1981)	= 0	All India
1975-76	Ray (1981)	< 0	Punjab

Out of the 15 studies listed in Table 3.1 above, in 8 studies the relationship is inverse. In another 7, it is not clear if there is any relationship.

Production Function[18]

We postulate a general production function[19]

$$X_0^{-p} = (A_1X_1)^{-p} + (A_2X_2)^{-p} \qquad \ldots (1)$$

where X_i (i = 0, 1, 2) refer to output, land and labour respectively.[20] The inputs X_i (i = 1, 2) are assumed to be measured in physical and quantity terms. A_i (i = 1, 2) are also treated as variables. It is suggested that they define land and labour inputs in efficiency or quality inputs as contrasted with X_i (i = 1, 2) which measure the same inputs in physical or quantity units. In other words, this production function takes into consideration the quality of the standard inputs.[21]

It will be noticed that it is a harmonic function with only one parameter p. p is considered a parameter of substitution. In other words (1) is a neo-classical function implying substitution between the two factors: land and labour. There are very few studies on land and labour substitution. Majority of these are based on Cobb Douglas type functions which implicitly assume an elasticity of substitution between land and labour equal to unity; a rather stringent assumption. Function (1), however, takes into consideration an elasticity of substitution which is different from 1.

Generally p is related to the elasticity of substitution concept. The exact relationship is:

$$\sigma = \frac{1}{1+p} \qquad \ldots (2)$$

where σ stands for elasticity of substitution. Thus as $p \to 0$ $\sigma \to 1$. However, p can take any values between –1 to + ∞ in view of the fact $0 < \sigma < \infty$.

Biological Technology Function

The basic idea of Biological Technical Change (BTC) is that the productivity of land can be increased through biological factors. More specifically, a *la'* Ruttan, *et al.*, "Increases in output per worker can also be achieved through increased land productivity, but only if the rate of increase in output per hectare exceeds the rate of change in the number of workers employed per hectare. It is useful to refer to technologies that increase output per hectare as biological technology".[22] In other words, BTC improves land productivity. However, to be defined as BTC proper the land productivity must be greater than the increases in labour-land ratio. If we write D for the operator d/dt, then this idea can be formulated as—

$$\left(\frac{DX_0}{X_0} - \frac{DX_1}{X_1}\right) - \left(\frac{DX_2}{X_2} - \frac{DX_1}{X_1}\right) > 0 \rightarrow BTC \qquad \ldots (3)$$

It will be noticed that this follows from the standard Solow[23] argument where he defined technical change as the difference between rate of growth of labour productivity and capital-labour ratio. It can be easily seen that (3) follows straight forwardly from Solow's argument. (3) can be written as—

$$\frac{DZ_{0,1}}{Z_{0,1}} - \frac{DX_{2,1}}{X_{2,1}} > 0 \rightarrow BTC \qquad \ldots (3')$$

where Z and X refer to output-input and input-input ratios respectively. The order of the subscripts is according to the numerator-denominators. Thus, $Z_{0\text{-}1}$ stands for (X_0/X_1) and $X_{2,1}$ for (X_2/X_1). The only difference between (3′) and original Solow relation is that in (3′) $X_{2,1}$ term is not weighted by share of labour in output. On the other hand, *BTC* in (3) and (3′) at the moment is a qualitative rather than a quantitative concept. All (3′) says is that if the rate of growth of land-productivity is greater than that of labour-land ratio, then there is *BTC*.

From (1) one can easily derive the growth equation

$$\frac{DZ_{0,1}}{Z_{0,1}} - \frac{DX_{2,1}}{X_{2,1}} = -S_1\frac{DX_{2,1}}{X_{2,1}} + S_1\frac{DA_1}{A_1} + S_2\frac{DA_2}{A_2} \qquad \ldots (4)$$

where S_i $(i = 1, 2)$ stand for the share of the input X_i and $\Sigma S_i = 1$. In (4) the left hand side is an in (3') and the right hand side can be interpreted as *BTC*. The difficulty in the definition of BTC by Ruttan, *et al.*, is that it is not sufficiently precise. Notice in (4) it includes the term S_1 $(DX_{2,1}/X_{2,1})$ i.e., the rate of growth of labour per unit of land. In other words, if one were to quantify the *BTC* as defined by Ruttan, *et al.*, it will depend upon the growth rate of inputs such as land and labour.

A more meaningful and precise way to define *BTC* is as follows:

$$\frac{DZ_{0,1}}{Z_{0,1}} - S_2\frac{DX_{2,1}}{X_{2,1}} = S_1\frac{DA_1}{A_1} + S_2\frac{DA_2}{A_2} \qquad \ldots (5)$$

so that the right hand terms are made up of qualitative or efficiency units only while the left hand term is based on the differences between the growth rates of productivity and labour intensity of land corrected by share of labour. It will be appreciated, and can be easily verified, that (5) follows from the production function in (1) and is quite general.

In our formulation A_i (i = 1, 2) are quality or efficiency units. There is no way one can observe these. Accordingly, one needs some hypothesis to explain the changes in these variables. For purposes of this paper we assume that A_2 is constant so that $DA_2 = 0$. Obviously this is a stringent condition. Our defense lies in two arguments. One, an appeal to convention. Both Cobb Douglas and CES production functions, which are commonly used, make such an assumption axiomatically. Two, the nature of the partial equilibrium analysis. Basically, we are assuming explicitly *ceteris paribus*; other factors remain the same. Our interest in this paper is somewhat limited. We wish to focus on BTC only. In a larger analysis, mechanical technical change is brought through A_2.[24]

BTC follows from a number of related activities such as high yielding varieties of seeds, facilities for controlled irrigation, fertilizers, etc. As Ruttan has rightly argued, *BTC* "shifts the fertilizer response surface, thereby facilitating the substitution of fertilizer for land."[25] In other words, the process of *BTC* may be captured through the use of fertilizers. Accordingly, we propose the hypothesis that A_1 depends upon the fertilizer-intensity of land; i.e., the more fertilizer that a farmer uses, the more productive the land becomes. After all, this, in a nutshell, is the basic assumption in all analysis of agricultural production. Fertilizers enrich land and make it more productive. There is ample literature on this idea and assumption. Practitioners of modern agriculture, and salesmen of chemical companies, swear by it. To give this idea some form we propose.[27]

$$\frac{DA_1}{A_1} = F\frac{DX_{3,1}}{X_{3,1}} \qquad \ldots (6)$$

where $X_{3,1} = (X_3/X_1)$ and X_3 stands for the quantity of fertilizers consumed. $X_{3,1}$ then, represents the fertilizer intensity of cropping. We have used the function F. There is no obvious form of F. General assumption are that $F' > 0$, i.e., increasing fertilizer does increase efficiency of cropping on land. Also $F' < 0$ in view of the law of diminishing returns and the idea of saturation. Since F states a relationship between the quality or efficiency of land and the level of fertilizer use, it may be called the Biological Technical Change Function (*BTC* Function).

In another sense (6) is a formulation of the idea that fertilizer is an input in agricultural production. A number of studies even use X_3 as a standard input in a Cobb-Douglas type production function. It can be easily verified that the substitution of (6) in (1) leads to the production function employed by Thirtle.[28] In that sense, (6) conforms to such ideas and formulations.

It is not easy to define a form for function *F*. As a first approximation we suggest a simple linear form: i.e.,

$$\frac{DA_1}{A_1} = \Upsilon \frac{DX_{3,1}}{X_{3,1}} \qquad \ldots (7)$$

One would expect $\Upsilon > 0$. However, it is quite possible, $\Upsilon < 0$, because of misapplication of this technology or because of side effects. Thus, $\Upsilon \lesseqgtr 0$.

Parameter Υ defines the elasticity of land quality with respect to fertilizer intensity. In simple language it may be interpreted as reflecting the effects of fertilizer use of land quality or efficiency.

Substituting (7) in (5) and assuming $DA_2 = 0$, we get

$$\frac{DZ_{0,1}}{Z_{0,1}} - S_2 \frac{DX_{2,1}}{X_{2,1}} = \Upsilon S_1 \frac{DX_{3,1}}{X_{3,1}} \qquad \ldots (8)$$

The right hand term is the measure of *BTC*. In our formulation $BTC = \Upsilon S_1 \frac{DX_{3,1}}{X_{3,1}}$. Thus, (8) provides a measure of *BTC*. It depends upon the rate of growth of fertilizer intensity, the share of land in the output and the elasticity of land quality with respect to fertilizer intensity. Basically, it follows from an equivalent relationship. However, if one may look for a causal relationship in an equivalence statement, one may argue that fertilizer use increases its intensity which effects the quality of land which in its turn effects productivity by extending further the isoquant.

Land-Productivity and Biological Technology

Our formulation so far consists of only two relations: the production function (1) and *BTC* function (7). Both these relations are structural. They do not include any economic argument. To make an economic model it is necessary to introduce economic considerations. To do so we postulate the relationship[29] between the price of fertilizers and the marginal product of fertilizers; namely,

$$\frac{Df_3}{f_3} = \frac{DP_3}{P_3} \qquad \ldots (9)$$

where f_2 refers to the partial of X_0 with respect to X_3. The implication of (9) is that fertilizer is like any input. The farmer or producer desires it not for its own sake but for its contributions to output or profit (9), thus is based on a derived demand suggesting that the farmer will be prepared to pay a price equivalent to its marginal product. The way we have formulated (9), it does not imply perfect market for fertilizer. It is consistent with a set of market imperfections. All (9) assumes is that the rate of growth of marginal product of fertilizer is equal to the rate of growth of its price.[30]

From (1), (7), and (9) we can easily derive the equation

$$Z_{0,1} = a_0 P_3^{a_1} X_{3,1}^{a_2} \qquad \ldots (10)$$

where $a_0 = A_{10}\, C$; $\qquad \ldots (11)$

A_{10} and C being constants of integration

$$a_1 = \frac{1}{(1+p)} \qquad \ldots (12)$$

$$a_2 = \frac{\Upsilon_p + 1}{1+p} \qquad \ldots (13)$$

(10) can be interpreted as an explanation of land-productivity in terms of fertilizer intensity and fertilizer prices as well as the elasticity of substitution between land and labour on the one hand and the impact of fertilizer intensity on land quality on the other. According to our interpretation fertilizer-intensity is the proxy for *BTC*. (10) then provides a formulation of the following theory. *BTC* affects the quality of land and extends further the frontiers of production. Fertilizer- intensity is a measure of *BTC*. These improvements in technology and land quality coupled with land-labour substitution affect the productivity of land.

Model and Identification

Our theoretical model may be written as follows:

$$X_0^{-p} = (A_1X_1)^{-p} + (A_2X_2)^{-p} \qquad \ldots (1)$$

$$A_1 = A_{10}\, X_{3,1}^{\Upsilon} \qquad (\bar{7}')^{31}$$

$$Z_{0,1} = a_0 P_3^{a_1} X_{3,1}^{a_2} \qquad \ldots (10)$$

These 3 equations contain six variable.[32] X_i (i = 0, 1, 2, 3), A_1 and P_3, X_0, X_1 and A_1 are indigenous while X_2, X_3 and P_3 are exogeneous. The parameters of interest are p and Y, which are contained in (10) indirectly. If we look at these equations we find that equation (10) is exactly identified[33]. Within the above system, (10) satisfies both order and rank conditions. Order conditions imply that at least two variables must be excluded. In case of (10) these excluded variables are A_1 and X_2. The rank condition is that the determinant B is non-singular, where B is

$$B = \begin{vmatrix} -p & -p \\ 0 & 1 \end{vmatrix}$$

Obviously, B will be singular only if p = 0. The rank condition thus, implies that p (≠) 0. It can be easily seen that the production function itself becomes degenerated when p = 0.[34]

In the above model, the only relation that can be quantified is (10). To estimate (10) one needs to introduce a stochastic term in it. For our empirical analysis, we assume that (10) is stochastic[35]. Since (10) is exactly identified, the method of least squares has all of the desired properties.

It will be noticed that (10) involves data on $Z_{0,1}$ —production and area under cultivation, p_3/P_0—prices of fertilizers and output, $X_{3,1}$—fertilizer intensity. These data are comparatively easy to obtain and are the most reliable. In contrast, the data on agricultural labour, machinery, capital, wage rates in agriculture, price of capital, are the most difficult, if not impossible, to obtain. Data on agricultural employment and capital investment are rarely available; whatever are available, are notoriously unreliable. The advantage of (10), thus, lies in the fact that it allows one to quantify *BTC* and land-labour substitution from comparatively reliable data.

Evidence on Technological and Structural Theories: A Hypothesis and a Reconciliation

It has been suggested that technology is of two kinds: mechanical and biological. The effect of both types of technologies is to push the production frontier further and thereby lead to increases in food production. In both literature, and various policy formulations at the government level, an implicit distinction is made between two types of technologies: (i) technologies that are neutral to the scale of operation; i.e. where $dZ_{0,1}/dFS = 0$; and (ii) those that are biased towards a larger scale of operation; $dZ_{0,1}/dFS>0$. It is generally assumed that biological technologies are scale neutral while mechanical technologies are biased in favour of large scale.[36] This assumption has never been tested.

The question is : is biological technology scale neutral? This is not a question of theory but a matter of fact. It can be answered only by an appeal to facts. An answer to this question involves the formulation of a relationship between farm size and biological technology. In the previous sections we have developed a parameter γ as a measure of biological technology. The question of the scale neutrality of biological technology translates into a relationship between γ and farm size. There are two relationships here: (a) biological technology is scale neutral; $d\Upsilon/dFS = 0$; and (b) biological technology is biased in favour of large farm size; $d\Upsilon/dFS>0$.

We have, thus, two hypothesis; (a) $d\Upsilon/dFS=0$ and (b) $d\Upsilon/dFS>0$.

Since we do not have any way to define farm size except to sort out various farms in discrete groups of small, medium and large, all that needs to be done to test these hypotheses is to estimate γ for different farm size.

These hypotheses have important implications in terms of interpreting evidence for the structural and technology schools. Assume that the evidence is in favour of the 'structural' arguments; i.e., small farms have larger land productivity; or $dZ_{0,1}/dFS<0$. Now introduce the factor of biological technology; γ. Let this technology be biased in favour of large farms; hypothesis (b); $d\Upsilon/dFS>0$. The effect will be that the land productivity in large farms will be higher so that the land productivity for small and large farms may not be different; $dZ_{0,1}/dFS = 0$. Thus, the seemingly contrary evidence in Table 3.1, (p. 64) may be reconciled by the hypothesis $d\Upsilon/dFS>0$.

Distribution of Farms by Size

To test the above hypotheses, we have obtained data on wheat farms in Punjab, India. These data are collected by Government of India through the Cost of Cultivation Surveys. These are the most extensive, and intensive, data sets ever collected by any agency anywhere in the world. These data are still unpublished. The government has released small parts of these data for researchers. The data was provided to us in a coded form so that the identity of individual farms would not be revealed. The data set we have relates to 149 wheat farms in 12 districts of Punjab for the year 1979.[37]

The analysis presented in this paper is based on five variables: quantities of output, land and fertilizers used, and prices of output and fertilizers. Data on quantities of output and land are generally the most reliable. In Panjab, the market for wheat is highly developed. The data on wheat prices, therefore, are also very reliable. Fertilizer is an industrial commodity and is generally distributed through the market. However, there is a governmental regulation on its distribution so that there is partly a black market in fertilizers. There is no well defined price of fertilizers since a part of it has to be imputed.[38]

In the literature on Indian agricultural studies, farm size has been divided into three categories; small, medium, and large.[39] In most studies, particularly on wheat, it is customary to define a small farm as one with less than 5 hectares and large farms as more than 10 hectares. In farm size studies <5 and >10 has become virtually a convention.[40]

On the above basis, the sample of 149 farms has been divided into small, medium and large farms. The break-up of this sample is given in Table 3.2.

TABLE 3.2 : Distribution of Farm by Size

Farm Size	*Definition (in hectares)*	*Number of Observations*
Small	< 5	55
Medium	> 5 and < 10	55
Large	> 10	39
	Total	149

Data and the Magnitude of the Variables

The data distinguishes between owned and leased land. Farm size is the land area, in hectares, actually cultivated by the farm. It has been obtained by the simple arithmetic sum as follows.

Farm Size = Land Owned + Land Leased in — Land Leased out. The land quality does vary over farms; though within narrow range. There is no measure of land quality. From the data that is available to us, it is not possible to develop a measure of land quality. It is precisely for this reason that the paper introduces the idea of biological technology effecting the quality of land.

The output data are straightforward. These refer to tonnes of wheat produced. The production data are the most reliable in view of the fact that these data have been collected for so many decades. There is a question about the quality differences in wheat. These differences are minimal. There is a sufficient uniformity in the varieties of seeds used and the quality of wheat produced in Punjab. The price of wheat is also quite standardized. The government fixes a procurement price below which the price cannot fall. This price is not far from the market price.[41] The data on wheat prices, therefore, are highly reliable.

The use of fertilizers in Punjab is now quite widespread.[42] Further more, the government is actively involved in the promotion and distribution of fertilizers. The data on fertilizer consumption, therefore, are easily available. However, in view of the governmental promotional and distributional policies, there may be certain biases in the consumption of fertilizers by small and large farmers. The data on consumption by small farmer may be biased upward while that of the larger farm downwards. On the other hand, small farmers would supplement their fertilizers with traditional measures, which would reduce this upward bias. The data we have are for the quantity of fertilizers actually used in production and not in terms of purchases by the farm. Since there are various types of fertilizers; ammonia, urea, etc. the data have been aggregated in NKP units.[43] This aggregation may reflcct the quality of fertilizer input. However, we have no information to assess this type of fertilizer quality.

The quality differences in fertilizer inputs and government distribution policies have a larger impact on the cost of fertilizers in terms of cultivation. The small farmers are likely to obtain the fertilizers at controlled prices and sell surplus quantities to larger farmers at higher prices. On the other hand, the larger farmers, in view of their social status, and better acquaintance with law are liable to obtain larger quantities at concessional prices. The larger farmers are liable to use fertilizers far more intensely than the observed statistics may show. Their relative costs may then be lower too. The data we have provide the information on the costs of fertilizers in terms of their use per hectare.

Table 3.3 below provides an idea about the magnitudes of the variables used in this analysis.

TABLE 3.3 : Output, Inputs and Prices by Farm Size (Average Values)

	Variable	*Small*	*Medium*	*Large*
A:	Quantities			
	$Z_{0,1}$ (tons)	25.60	28.46	28.26
	$X_{3,1}$	126.40	161.34	164.24
	X_1 (hectares)	3.23	7.38	14.47
B:	Prices			
	P_0 (rupees per tonne)	114.21	114.57	114.67
	P_3 (rupees per hectare)	8.65	19.07	36.93

The large farms are virtually 4.5 times the size of small farms. They do use fertilizers more intensively and the cost of fertilizers for large farms is considerably greater, as one would expect.[44] The table suggests, quite clearly, that the large farms use modern technology heavily so that in their case productivity may be explained by the arguments of the technology school.[45]

The land productivity for all farm sizes is quite similar. One may, thus, conclude that $dZ_{0,1}/dFS = 0$. This would confirm the evidence of other studies[46] and the arguments of technology school. This observation would not, a priori, support the structural school.

Farm Size and Biological Technology

Given the data on all variables, it is possible to estimate γ for all the three farm sizes and test the hypothesis: $d\Upsilon/dFS>0$. γ can be estimated from equation (10) by indirect least squares method. Accordingly, we have estimated equation (10); namely,

$$Z_{0,1} = a_0 P_3^{a_1} X_{3,1}^{a_2} \qquad \ldots (10)$$

for all the three farm sizes. Table 3.4 below provides the estimates of the coefficients of (10).

TABLE 3.4 : Estimates of Coefficients

Farm Size	*n*	a_0	a_1	a_2	$\overline{R}^2$
Small	55	3.125	.351	.260	.25
		(16.599)	(3.951)	(4.265)	
Medium	55	2.858	.265	.286	.33
		(20.763)	(3.124)	(5.293)	
Large	39	2.465	.273	.428	.33
		(12.914)	(2.546)	(4.268)	

Note : Figures in brackets refer to t–ratios.

These are cross section data. Generally, $\overline{R}^2$ values for cross section data are not high. The $\overline{R}^2$ values in Table 3.5; therefore, are reasonable. All the coefficients a_i (i = 0, 1, 2) are significantly different from zero. The standard errors are low, and t-ratios are all greater than 2. The statistical results, thus, are quite good. These are acceptable by any statistical standard.

Looking at the coefficients a_i (i = 1, 2), we note that both a_i are significantly different from zero for all the three farm size groups. Our theory requires that a_2 be also different from 1.[47] The estimates of a_2 are different from 1 also. The estimates of a_i (i = 1, 2) are thus consistent with our theory.

From estimates of a_i (i = 1, 2), structural parameters σ and γ can be estimated from the following relations.

$$\sigma = a_1 \,;\; \rho = \left(\frac{1}{a^1} - 1\right) \qquad \ldots (12')$$

$$\Upsilon = \frac{1}{\rho}\left(\frac{a_2}{a_1} - 1\right) \qquad \ldots (13')$$

Table 3.5 below gives the estimates of σ and γ for different farm sizes.

TABLE 3.5 : σ and ϒ by Farm Size[48]

Farm Size	σ	ϒ
Small	.35	–.14
Medium	.26	.03
Large	.27	.20

The estimates of σ are low for all the three farm sizes. This is what should be expected. Land and labour are complimentary factors and not substitutes. These low values are consistent with other studies. Hans

Binswanger has estimated γ for U.S. via translog cost function. He estimates its value at 0.024.[49] Yeung and Roe have estimated this coefficient for Japan by CES function and the value they find turns out to be .3.[50] Diwan and Kallianpur have estimated σ for wheat and rice crops for time series data in India. Their estimates are .37 and .24 respectively for wheat and rice.[51] These values of σ estimated by different authors, by different methods, for different sets of data and for different countries are unusually close.[52] One does not expect on either a priori or theoretical grounds for σ to be different for different farm size. There is no expectation for σ to vary over farm size. The values of σ in Table 3.5 above confirm this expectation.

Υ is a new parameter that we have introduced in our analysis. No estimates about its magnitude are available except our own study based on all India time series data for rice and wheat. The estimates of Υ for rice in this study is –.017 and for wheat .06.[53] The estimates in Υ in Table 3.5 above are fully consistent with these earlier estimates. Considering that the data sets for these two studies are very different one is time series and the other cross section, one is aggregate while other micro data, the consistency of these estimates is remarkable. What is still more striking is that the estimates of γ for rice in the previous study and for small farms in this is negative. This is consistent because majority of rice production in India is carried on in small farms.[54]

We can now test our hypothesis. The estimates of Υ for farm size clearly support the hypothesis that $d\Upsilon/dFS>0$. In other words, biological technology is not scale neutral. It is instead biased in favour of large farms. The conclusion is consistent with both recent studies: Deolalikar (1981)[55] and Rao and Chotigeat (1981).[56]

Conclusions and Policy Implications

In earlier studies researchers found the relationship between land productivity ($Z_{0,1}$) and farm size (FS) to be inverse, i.e., $dZ_{0,1}/dFS>0$. In popular terms, small farms are more productive.[57] More recent evidence suggests that there is no relationship between productivity and scale: $dZ_{0,1}/dFS = 0$. These two contrary pieces of evidence have generated a controversy on the relationship between productivity and scale.

This paper reconciles this seemingly contrary evidence. It argues that the inverse relationship is consistent with the evidence of no relationship if scale-bias of technological inputs is taken into consideration. Thus, if (i) biological technology (Υ) is biased in favour of large farms; $d\Upsilon/dFS>0$, and (ii) the invesre relationship holds: $dZ_{0,1}/dFS<0$, then one will observe $dZ_{0,1}/dFS>0$. The reconciliation of the seemingly contrary evidence, then, lies in the scale-biasedness of technology. The testable hypothesis is: is $d\Upsilon/dFS>0$? The paper tests this hypothesis for a new set of cost of cultivation data. The finding is that $d\Upsilon/dFS>0$.

The conclusion of this paper is that the thesis of an inverse relationship between land productivity and farm size is valid and is consistent with the evidence from recent empirical studies.[58] This conclusion has important implications.

It raises doubts about the scale neutrality of biological technology inputs. It seems there is a contrariness between agronomy and economic studies. Scale-neutrality hypothesis follows from the agronomy studies conducted in experimental farms. By abstracting socio-economic relations these studies may not be able to capture some of the implied biases of biological technology inputs. Public and economic policies based on agronomy studies may not be appropriate.

The validity of the inverse relationship points to some meaningful research questions about (a) the labour quality and (b) the theory of work motivation. It seems that small farms are able to increase productivity by substituting labour for technology. The question is: what are the elements of this substitution? *Prima facie,* such substitution must involve qualitative dimensions of labour inputs. What are the determinants of labour quality? Literature has concentrated on human capital type ideas. Human capital is complimentary with and not a substitute for technology. One needs to look into other facets of labour quality.[59] Again small farms are able to obtain both greater amounts of quantitative and higher levels of qualitative labour input. How and why? This involves a better understanding of factors responsible for work motivation.

In the last two decades, the international institutions and governments have concentrated their attention and policies on the provision of new technological inputs. It is now a conventional wisdom that provision of technological inputs is the only desirable and meaningful avenue for productivity growth. In a way it has been an easier policy because it has affected social relations implicitly and not offended the political base of governments. The inverse relationship brings into focus new options for productivity growth and related policies.

The acceptance of this thesis favours different, and difficult, policies dealing with land ceiling, land distribution; land legislation and land reforms. These are exactly the policies that governments have shield away from.[60] These policies involve major social change.

The impact of these general issues on purely economic policies is quite large. Virtually, every developing country is suffering from the problem of severe unemployment. The economic policies in terms of provision of technological inputs has aggravated the unemployment situation.[61] Economic policies following from the inverse relationship, on the other hand, contribute both to the solution of production and unemployment problem. In turn these have major consequences in terms of price support, import and export policies. This thesis provides an argument for alternative development process and policies.[62]

Note : Reproduced from *Indian Journal of Quantitative Economics,* Vol. I, No. 1 (1985), Punjab School of Economics, G.N.D. University, Amritsar (India).

NOTES

1. See Lappe' and Collins, (1977).
2. World Bank, in a generally conservative estimate, estimates that several million people are undernourished. In its own words, ''Nevertheless, allowing a wide margin for uncertainties, (in the definitions of undernourishment) the evidence is strong enough to conclude that several hundred million people are undernourished," World Bank, (1980, p. 59). Undernourishment refers to shortage of food and not an imbalance between calories and proteins.
3. It is not easy to develop a concensus about how to define poverty. However, the povery is so widespread and absolute that all definitions portray the same reality. According to World Bank, ''The inclusion of China thus suggests a global total of close to one billion people living in absolute poverty'', World Bank, (1980), p. 78. (emphasis added). One billion people mean one fourth of the total world population.
4. ''The *1980 World Development Report* and the FAO's Study of future agricultural growth prospects, *Agriculture: Towards, 2000,* were both completed when the world economy was more buoyant than it is now. Nevertheless, both concluded that neither poverty nor undernutrition could be eliminated in this century, even under the most favourable assumptions about agricultural development and overall economic growth," World Bank, (1982), p. 86.
5. ''FAO reckons that only 10 to 15 per cent of unused arable land in 1980 might

be cultivated by 2000," World Bank, (1982). p. 59.

6. See Schultz, (1964).
7. If one reads the literature on development and agricultural production in early 1940's, one finds that the conventional wisdom then was that the farmers in poor countries were both irrational and inefficient. Adjectives used for these farmers were; backward, superstitious, unscientific, etc.
8. It is such evidence that has given a push to the establishment of international research institutes all over the world such as CLMMT, IFPRI, etc.
9. See, among others, Binswanger and Ruttan, (1978), Evenson, (1975), Hayami and Ruttan, (1971).
10. Police of the government and international agencies are based on it. Evidence is sought, easily obtained, and categorically stated; no ifs, buts, qualifying clauses, etc. World Bank is the greatest promoter of this thesis. Their language speaks for itself. 'The average productivity of cultivated land has been growing at about 2.2 per cent a year for the past 20 years... These productivity gains have been achieved largely by improving the availability and reliability of irrigation and increasing the use of new seeds and fertilizers," World Bank (1982), p. 61.
11. See Rutan, *et al.*, (1978), Thirtle, (1981), and Kallianpur, (1983).
12. It is interesting to note that much of the agronomy studies assume the availability of quality labour. One wonders what would happen to the conclusion of these studies if they had to depend upon less quality labour.
13. This involves various issues in the 'theory of motivation.' Galbraith, (1967) classifies motivation under four categories: (i) compulsion, (ii) pecuriary, (iii) identification and (iv) adaptation, See Chapter XI and XII. Actually, his categories (i) and (ii) involve compulsion and (iii) and (iv) involve voluntary involvement.
14. There is ample literature, and evidence, on the low land productivity in Zamindari system in India. Genovese, (1961) also makes the same argument in case of the slavery system.
15. It will be appreciated that "identification" is one of the most important motive forces. it is this identification that lies at the heart of "small is beautiful".
16. See Galbraith, (1967), Chapter XI and XII. For an illuminating discussion on the nature of compulsion in wage payments, see Illich, (1981).
17. It needs to be added that a number of studies have derived this hypothesis from other arguments. However, these arguments are quite consistent with the summary arguments presented in the text.
18. This and the subsequent three sections are taken from Diwan and Kallianpur, (1983). The present section has been adapted from Diwan, (1971).
19. This form has been used by Sato and Beckman, (1968) and Diwan, (1970), Diwan has shown that if X_1 refers to capital and A_2 is defined as a function of capital intensity of labour, then (1) becomes a VES (Variable Elasticity of

Substitution) production function.—The derivation of Cobb Douglas and Leontief function follow easily.

20. Agricultural economists have, no doubt, assumed a production function that involves land and labour as inputs. General studies in the economics of growth and production, on the other hand, have assumed land as being given. These have concentrated on the inputs capital and labour. In this respect there are differences between the standard neo-classical production function and agricultural production. All the same agricultural production functions have some of the properties assumed for neo-classical functions such as convexity, substitution, etc.
21. It is now becoming quite fashionable to define general production functions to take into consideration "qualitative" dimensions of inputs. Translog functions are built on these considerations.
22. Ruttan *et al*, (1978), p. 46.
23. In Solow (1957)

$$\frac{DA}{A} = \frac{Dq}{q} - W_k \frac{Dk}{k}$$

where A is technical change, q is output-labour ratio and k is capital-labour ratio; W_k is the share of capital in output. Solow's argument involves a number of simplifications and resulting biases. For example, see Diwan, (1968).

24. See Kallianpur, (1983).
25. Private communication from Professor Vernon Ruttan, University of Minnesota.
26. To cite the most recent example, John Antle, (1983) states, "The fertilizer variable is used here as a proxy for 'modern' inputs." This is all the justification he provides for the inclusion of fertilizer as a separate variable in a Cobb Douglas type of production function.
27. Diwan, (1971) uses a similar function to analyze the impact of education on labour efficiency.
28. Thirtle, (1983) uses the production function

$$X_o^{-a} = (A_{10} X_1^{b1} X_3^{b3} e^{bt}) + (A_{20} X_2^{b2} X_4^{b4} e^{b't})$$

i.e., $X_0^{-a} = Y_1 + Y_2$

It can be shown that Y_1 follows if we assume $A_1 = A_{10} X_{3,1}^{\sigma}$, in our equation (7).

29. It will be noticed that our postulate implies a sort of marginal productivity theory and all that goes with it.
30. Strictly one may question whether f_3 can be defined. In case we argue that X_3 affects X_0 through A_1 and hence X_1, then $f_3 = f_1\ (\partial X_1/\partial X_3)$. It can then be affected by the last term on the right hand side. The argument in the text is quite defensible; the relevant conditions can

be satisfied.

31. It will be noticed that (7) and (7′) are equivalent because (7′) is an integral of (7) overtime.
32. We have assumed that A_2 is a constant.
33. It will be noticed that (1) is also exactly identified. The rank conditions are that $\Upsilon \neq 0$ and $^{p} \neq -1$. (7′), on the other hand, is over identified.
34. It may be observed that this is a rather stringent condition. If one refers to the production function literature one will notice that CES and VES functions do imply a nonzero value for ^{p}.
35. There are two ways of doing so. One is to argue that the original relations (1), (7), and (8) are themselves stochastic relations. If so, then the substitution and manipulation of the three relations into (10) will bring an additional term which will be a combination of the stochastic terms in the three relations. The second approach is to introduce the stochastic term in (10). The implication is that the model holds exactly but not the marginal relation. Statistically, both approaches lead to the same form. The implications, no doubt, are different.
36. See, for example, World Bank, (1982), Ch. 6, Leaf, (1983), Blynn, (1983).
37. We have data for all the important inputs: land, labour, capital, bullocks, fertilizers as well as for their prices. However, some of the quantities and virtually all the prices involve a number of imputation and assumptions.
38. Srivastava, (1983) argues that the small farmers may be selling their quota of fertilizers to larger farmers.
39. See the references to the studies in Table 3.1.
40. It needs to be added that this is strictly in the context of India where land ceiling legislation was passed immediately after the independence in 1947. In other counties, the concept of small and large may be much different.
41. There is an elaborate process by which the procurement prices of wheat are determined by the Agricultural Prices Commission. The wheat lobby is quite strong and the question regarding the price are settled at Central Cabinet level.
42. The "green revolution" was launched in Punjab in mid 1960s. It has been called as successful experiment. There is now a vast literature on it. For example, see Blyn, (1983), Leaf, (1983), and Frankel, (1971).
43. The details on this aggregation are not available. It is generally believed that it is 5-10-5 mixture.
44. In Panjab they have the necessary facilities in terms of tubewells, for an intensive use of fertilizers, see Dhawan, (1982).
45. Much of the lobbying for modern technology comes from larger farms. These are also the darlings of the government and international agencies for using "modern" methods of production.
46. See, for example, Deolaliker, (1981).
47. It needs to be emphasized that this is a rather stringent condition following from the rank conditions for identification.

48. The standard errors of σ and Υ have not been estimated. These can be estimated by the first approximation of Taylor expansion. In view of the fact that the coefficients a_i $(i = 1, 2)$ are highly significant, the likelihood is that the estimates of σ and Υ are also significant; σ is obviously so.
49. Binswanger, (1978), p. 239.
50. Yeung and Roe, (1978), p. 251.
51. Diwan and Kallianpur, (1983).
52. The closeness of these values lends credibility to our theoretical formulation, methods of estimation and data set.
53. Diwan and Kallianpur, (1983).
54. One of the major rice producing areas in India is made up of Andhra Pradesh and Tamil Nadu. The largest farms in rice production are close to 5 hecatares; a farm size which we have termed small in this study.
55. To quote "In this paper, regional data from India have been used to test the hypotheses that (a) the small farm sector as a whole enjoys higher yields per unit of land than the large farm sector in Indian Agriculture, but (b) this yield advantage of the small farm sector diminishes, and in fact even reverses, with technical change (primarily of the Green Revolution type) in agriculture. The data seems to confirm both the hypothesis," Deolaliker, (1981), p. 278.
56. "The results suggest that larger sized land holdings are not less productive in all circumstances, as may be generally believed. If hired labour is employed in preference to family labour and if more nontraditional capital as opposed to traditional capital is used, larger sized holdings and higher productivity can go together," Rao and Chotigeat, (1981), p. 574.
57. For some anecdotes, see World Bank, (1980. p. 42.
58. This is consistent with the findings of Blyn, (1983). "All evidence examined above indicated that gross output value per hectare decreased with holding size" (p. 719; emphasis added). It is for this reason he has argued that the land ceiling be raised to 25 acres. In this opinion 10 acres is too low.
59. Some of these issues are hinted in Diwan, (1982).
60. The issue of land reform in El Salvador is so important that U.S. government has, at least on paper, urged the El Salvador Government to incorporate in its policies. In other words, it can contribute to revolutionary politics.
61. See Aggarwal, (1979) for the negative impact of tractorization on unemployment.
62. For some elements on an alternative development, see Diwan and Livingston, (1979).

REFERENCES

Aggaral, Bina, (1979), "Tractorisation, Productivity and Employment: A Reassessment," *Journal of Development Studies,* Vol. 16.

Antle, John M., (1983), "Infrastructure and Aggregate Agricultural Productivity:

International Evidence," *Economic Development and Cultural Change*, pp. 609-619.

Bardhan, Pranab, (1973), "Size, Productivity and Returns to Scale: An Analysis of Farm Level Data in Indian Agriculture," *Journal of Political Economy*, pp. 1370-86.

Bharadwaj, K., (1974), "Notes on Farm Size and Productivity," *Economic and Political Weekly*, Vol. 9, No. 13, pp. All-A24.

——, (1974b), *Production Conditions in Indian Agriculture: A Study Based on Farm Managed Surveys*, Cambridge University Press.

Bhattacharya, N. and G.R. Saini, (1972), "Farm Size and Productvity: A Fresh Look," *Economic and Political Weekly*, pp. A63-A72.

Binswanger, Hans, (1978), "Measured Biases of Technical Change: The United States," pp. 215-243, in *Induced Innovation: Technology, Institutions and Development*, John Hopkins University Press.

—— and V.W. Ruttan, (1978), *Induced Innovation: Technology, Institutions and Development*, John Hopkins University Press.

Blyn, George, (1983), "The Green Revolution Revisited," *Economic Development and Cultural Change*, July, pp. 705-726.

Chattopadhyay, M. and Ashok Rudra, (1976), "Size-Producitivity Revisited," *Economic and Political Weekly*, Vol. 11, No. 39.

Deolaliker, Anil B., (1981), "The Inverse Relationship Between Productivity and Farm Size: A Test Using Regional Data from India," *American Journal of Agricultural Economics*, pp. 275-279.

Dhawan, Banarsi, (1982), *Development of Tubewell Irrigation in India*, New Delhi: Agricole Publishing Co.

Diwan, Romesh K, (1982), "Economics of Love or An Attempt at Gandhian Economics," *Journal of Economic Issues*, Vol. XVI, No. 2, June 1982, pp 413-433.

——and Dennis Livingston (1979), *Alternative Development Strategies and Appropriate Technology: Science Policy for an Equitable World Order*, Elmsford, New York: Pergamon Press.

——, (1971), "Impact of Education on Labor Efficiency," *Applied Economics*, pp. 127-135.

——, (1970), "About the Growth Path of Firms," *American Economic Review*, Vol. 50, No. 1, pp. 30-43.

——, (1968), "A Bias in the Measurement of Technical change," *Journal of Financial and Quantiative Analysis*, pp. 471-477.

——and Renu Kallianpur (1983), "Biological Technology and Land Productivity: Fertilizer and Food Production in India," *Paper presented at the Second World Congress of Land Policy*, Boston, June 20-24.

Frankel, F.R. (1971), *India's Green Revolution*, Princeton, New Jersey: Princeton University Press.

Galbraith, John Kenneth, (1967), *The New Industrial State*, Boston, MA: Houghton Miflin Co.

Genovese, Eugene D. (1961), *The Political Economy of Slavery*, New York: Vintage Books.

Ghose, Ajit Kumar, (1976), "Farm Size and Land Productivity in Indian Agriculture: A Reappraisal, *Journal of Development Studies*, Vol. 16, pp. 27-49.

Hanumantha Rao, C.H., (1975), *Technological Change and Distribution of Gains in Indian Agriculture*, Studies in Economic Growth, No. 17, New Delhi: Institute of Economic Growth.

——, (1968), "Farm Size and the Yield Per Acre: A Comment," *Economic and Political Weekly,* Vol. 3, No. 37, pp. 1413-1414.

Illich, Ivan, (1981), *Shadow Work,* Salam, NJ: Marion Boyers, Inc.

Kallianpur, Renu (1983), *Productivity and Technical Change: Empirical Evidence Analysis of Indian, U.S. and Japanese Agriculture,* Ph.D. Thesis (Managerial Economics), Rensselaer Polytechnic Institute.

Khan, M.H. and D.R. Maki, (1979), "Effects of Farm Size on Economic Efficiency: A Case Study of Pakistan," *American Journal of Agricultural Economics,* Vol. 61, No. 1.

Lappe', Frances Moore and Joseph Collins, (1978), *Food First: Beyond the Myth of Scarcity,* San Francisco: Institute for Food and Development Policy.

Leaf, Murray J., (1983), "The Green Revolution and Cultural Change in Punjab Village: 1965-78," *Economic Development and Cultural Change,* January, pp. 227-270.

Mauzumdar, D., (1965), "Size of Farm and Productivity: A Problem of Indian Peasant Agriculture," *Economica,* pp. 161-73.

Rao, A.P., (1967), "Size of Holdings and Productivity," *Economic and Political Weekly,* Vol. 2, No. 44, pp. 1989-1991.

Rao, Vaman and Tosporn Chotigeat, (1981), "The Inverse Relationship Between Size of Land Holdings and Agricultural Productivity," *American Journal of Agricultural Economics,* Vol. 63, No. 3, pp. 571-574.

Ray, S., (1981), *Policy Planning for Agricultural Development: The Indian Example,* Boulder, Colorado: Westview Press.

Rudra, Ashok, (1968a), "Farm Size and Yield Per Acre," *Economic and Political Weekly,* pp. 1041-44.

——, (1968b), "More on Returns to Scale in Indian Agriculture," *Economic and Political Weekly,* Vol. 3, No. 39, pp. A33-39.

Ruttan, V.W., H.P. Binswanger., Y. Hayami, W.W. Wade, and A. Weber, (1978), "Factor Productivity and Growth: A Historical, Interpretation," in *Induced Innovation,* pp. 44-87.

Saini, G.R., (1971), "Holding Size, Productivity and Some Related Aspects of Indian Agriculture," *Economic and Political Weekly,* June, pp. A79-85.

Sato, R. and Beckman, M.J., (1968), "Neutral Inventions and Production Functions," *Review of Economic Studies,* pp. 57-65.

Schultz, T.W., (1964), *Tranforming Traditional Agriculture,* New Haven, CT: Yale University Press.

Sen, A.K. (1962), "An Aspect of Indian Agriculrure," *Economic Weekly,* pp. 243-46.

——, (1964), "Size of Holding and Productivity," *Economic Weekly,* pp. 323-26.

——, (1975), *Employment, Technology and Development,* Oxford: Clarendon Press.

Solow, R.M., (1957), 'Technical Change and Aggregate Production Function," *Review of Economics and Statistics,* Vol. 39, pp. 312-20.

Srivastava, A.P., (1983), "Agricultural Production and Increasing Fertilizer Application," *Economic and Political Weekly,* March 26, pp. A17-21.

Thirtle, Colin G., (1983), "A Reformulation of Induced Innovation Model," *Economic Development and Cultural Change,* forthcoming.

——, (1983b), "The Specification of Agricultural Production Functions: A Note," *Journal of Development Economics,* forthcoming.

World Bank, (1980), *World Development Report 1980,* New York : Oxford University Press.

Yamada, S. and V.M. Ruttan, (1980), "International Comparisons of Productivity in Agriculture," *New Developments in Productivity Measurement and Analysis.* W. Kendrick and N. Vaccara (eds.), Chicago : University of Chicago Press, pp. 509-585.

Yeung, Patrick and Terry L. Roe, (1978), "A CES Test of Induced Technical Change : Japan," in *Induced Innovations,* pp. 243-260.

4

TRACTORS, TUBEWELLS AND CROPPING INTENSITY IN THE INDIAN PUNJAB

BINA AGARWAL

I. INTRODUCTION

The effect of tractors and tubewells on cropping intensity can have a crucial bearing on what impact they have on the farm's annual output and employment per hectare. Any crop-specific positive effects on output, for instance, could be enhanced, or detrimental employment effects neutralised, through the cropping intensity factor (see Agarwal, 1980). It is in this context that arguments in support of tractorisation often draw heavily on the fact that tractors could help to increase cropping intensity, by enabling the farmer to save time and hence grow an extra new crop, or to devote more area to existing crops. Likewise, an important advantage of tubewell irrigation over canals is seen to lie in its ability to promote more intensive cropping by making available an assured water supply throughout the year, and by enabling closer supervision over the quantum and timing of irrigation. These would be important factors influencing the farmer's decision regarding both what to grow and how much land to plant in a season.

While there is a fair amount of support in the literature for the contention that tubewells add to cropping intensity (Billings and Singh, 1969; Kaneda, 1969; McInerney and Donaldson, 1975; NCAER, 1973; Rao, 1975; Vashishtha, 1975), the evidence for tractors varies: in some previous studies, tractors have been found to add to cropping intensity (Chopra, 1972; Johl, 1970; Grewal and Kahlon, 1972; Motilal, 1973), and in others to have a neutral effect (Rao, 1975; Vashishtha, 1975;

Billings and Singh, 1969). Reservations may, however, be expressed with respect to the accuracy of both sets of studies dealing with the tractor effect.[1]

Firstly, many of them do not take account of factors such as the source of irrigation, the level of irrigation, farm size and agro-climatic zones, which could also significantly affect cropping intensity. Hence some of the effects attributed to a tractor may in fact be unrelated to it.

Secondly, most studies use the ownership of a tractor as a proxy for usage: farms where tractors are hired in are counted as bullock farms. This is likely not merely to bias the results but also to obscure any differences in the impact of owned tractors relative to hired ones. For instance, a tractor owner would have greater flexibility in its use and hence be able to better reap any advantages than would a tractor hirer, especially where tractors are hired from other farmers (as is usual in the Punjab, where many of the Indian studies are based) rather than through an outside agency. Normally the owner will hire out his machine only after fulfilling his own requirements. The farmer hiring in the tractor would have to adjust the timings of his operations accordingly, thus losing some of the 'timeliness' advantage. This, plus the uncertainty regarding the availability of the tractor as and when he requires it over the year, can affect his cropping decisions.

In the studies mentioned, one or more of these aspects have been neglected. Johl, for instance, does not separate out the effect of any of the specified factors. Chopra and Motilal take account only of farm size. Grewal and Kahlon adjust for farm size and their farms are uniform in their source of irrigation, but as in other studies, so in theirs, tractor hiring is not explicitly taken into account. This last point also constitutes my main reservation as regards Rao's and Vashishtha's results. In both these studies, adjustment has been made for farm size, per cent area irrigated and the source of irrigation, but the computation of the tractor effect is based on a comparison of tractor owners with non-owners (irrespective of hiring in).

Thirdly, in all the noted studies, cropping intensity is defined conventionally as the ratio (expressed as a percentage) of gross cropped area to net sown area.[2] While this index is straightforward to interpret, it suffers from one major weakness: it takes no account of the differences between farmers in their ability to intensify cultivation further. If the farms compared have the same crop rotations over the year, the index presents no problem. Even when they have differing rotations the comparison is valid as long as no differences exist in the time taken by the alternative crops to mature. A wheat-paddy rotation would thus be comparable to a wheat-groundnut rotation since both paddy and groundnut take roughly four and a half months. However, when farmers are growing crops with different maturing periods, erroneous conclusions may be drawn from this index. Consider, for instance, two farmers both with five hectares of net sown area, one having a wheat-paddy rotation and growing five hectares of wheat in *rabi* (broadly mid-October to end of April) and three hectares of paddy in *kharif* (broadly May to mid-October), and the other growing only sugarcane (a 10-12 month crop) on all five hectares. By the conventional index, the first farmer will have a cropping intensity of 160 per cent and the second of 100 per cent. This is despite the first having two hectares idle for half the year and the second having the entire area occupied throughout the year. That the first farmer has some discretion in intensifying cultivation and the second has little, is obscured by the index. In fact, from the above results, we would conclude just the opposite.

In short, the main lacunae in previous studies may be traced to :

(i) their failure to separate the effects of tractors or tubewells from the impact of other variables; and/or
(ii) their basing their conclusions about the tractor effect on a comparison of bullock farms with tractor-owning farms, without taking account of tractor hiring on bullock farms; and
(iii) their basing their analysis only on the conventional index of cropping intensity.

The present study attempts to take care of all three aspects and so seeks to capture the cropping intensity effects of tractors and tubewells more accurately. Regarding the third aspect, an alternative to the conventional index is specified here, taking account of the differences in the time-duration of crops grown on different farms over the year. Both the conventional and the alternative indices have been used in the present analysis.

In the following sections of the paper, Section II gives the specification of the alternative cropping intensity index; Section III provides details of the data, the hypotheses and the methodology used for the empirical analysis; Section IV presents the results; and Section V contains a summary of the main findings and some concluding comments.

II. SPECIFICATION OF AN ALTERNATIVE CROPPING INTENSITY INDEX

The alternative index has been specified as follows :

$$\frac{\sum_{i=1}^{n} a_i t_i}{12 A_s} \times 100$$

Here a_i = area under crop '*i*'; t_i = average duration (in months) of crop 'i'; A_s = net sown area. The maximum value for this index is 100, that is, when every cultivated hectare of the farm is occupied for all months of the year. This index thus helps to compute the 'hectare-months' for which farm land is unoccupied and hence to determine the scope that exists for further increasing cropping intensity, given the farmer's choice of cropping pattern.

Crops grown under 'mixed cropping', that is, more than one crop being grown on the same plot of land at the same time (as, for instance, wheat-gram), have been counted as a single crop, with the total time for which a plot of land is occupied in this way being taken as the time duration of the 'mixed crop'. (In the present sample such cases were very few.) I shall henceforth be referring to the cropping intensity index computed by the conventional method as the 'conventional index' (I_c) and the one specified here as the 'crop-duration index' (I_d).

III. DATA, HYPOTHESES AND METHODOLOGY

Data

The data used relate to a sample of 237 farms[3] taken from the principal wheat-growing areas of the Indian Punjab, and covering 20 tehsils spread across all the districts of the state. The data are for the crop-year 1971-72, and were collected under the 'Comprehensive Scheme for Studying the Cost of Cultivation of Principal Crops' by the Punjab Agricultural University, for the Directorate of Economics and Statistics, New Delhi. A three-stage stratified random sampling was undertaken, with the tehsil as the primary unit, a cluster of three villages in each tehsil as the secondary unit, and the farms in each cluster as the final unit. Information was obtained on the basis of day-to-day observation of selected cultivators by full-time researchers residing in the villages.

This method is more accurate than the commonly used 'recall method' which relies on the memory of the respondent.

All the sample farmers own at least a part of the land they operate. The farms are of differing sizes ranging from 0.8 hectares to 33.7 hectares with the bulk (78.5 per cent) being over four hectares in size; and are spread over three agro-climatic zones. Each zone is broadly homogeneous with respect to cropping pattern, climate and level of rainfall, but differs from each of the other zones on these counts.[4]

Hypotheses

On a *priori* grounds (for reasons already stated) we would expect cropping intensity to be positively related to the following factors: the use of tractors in comparison with bullocks; irrigating with a tubewell rather than relying on canals alone; ownership of a tractor rather than relying on a hired one; and the percentage of gross cropped area irrigated. Further, *ceteris paribus,* we would expect cropping intensity to be negatively related to farm size. This is an important relationship and needs elaboration. Among reasons for expecting this inverse relationship are the following:

(i) *A managerial constraint* : The greater the area cultivated, the more would be the requirements of labour supervision and management, particularly under labour-intensive irrigated farming. This would set a limit on the amount of land the farmer decides to plant in a season. Also, land available between the main crop seasons is less likely to be planted with minor crops the larger the farm, since additional supervision would be needed both for the minor crops and for ensuring the timely preparation of land for the subsequent main crop. At what level a managerial constraint is encountered would depend on the availability of reliable persons (usually family members and/or permanent labourers) for supervision;

(ii) *A power constraint* : The larger the farm the greater is the likelihood of the farmer having to face this constraint. Even a farmer owning a tractor or a tubewell can encounter this problem if the farm exceeds the size for which a single tractor and/or tubewell would suffice, and yet is not large enough to make the purchase of an extra one worthwhile. This limit could again vary, depending on the availability of supplementary means (such as bullocks for tillage and canal water for irrigation) and/or the ability of the farmer to hire in such supplementary services;

(iii) *Specific circumstances such as poor quality, inadequately levelled or temporarily waterlogged land* : The larger the farmer the more likely is he to leave such land fallow, at least for a part of the year, while the smaller farmer might put in the extra effort needed to cultivate it.

Finally, we would expect differences in soil and climatic conditions (for which the agro-climatic zones are a proxy) to affect cropping intensity through their impact on cropping patterns.

Methodology

The sample farms have been divides broadly into the following technique categories:

(i) *Bullock farms* : those ploughed exclusively with bullocks;

(ii) *Tractor farms* : those ploughed[5] with tractors (owned or hired in). Here, in addition to tractors, bullocks might also be used for ploughing on a part of the cultivated area. The tractor-using farms are divided further into (a) tractor-owning farms, i.e. those where owned tractors are used (although, additionally, tractor services may be hired in), and (b) tractor-hiring farms, i.e. those where no tractor is owned but tractor services are hired in for ploughing;

(iii) *Canal farms* : those exclusively canal irrigated;

(iv) *Tubewell farms* : those irrigated by tubewells (this category includes pump-sets). Here canals may be used to supplement tubewells. In 97 per cent of the tubewell farms, the tubewells used are owned. The average cropping intensities of farms classified by combinations of the above technique categories have been tabulated.

The relative effect on cropping intensity of each of the factors mentioned in the hypotheses has been measured through multiple regression. The variables entering the regression equation are as follows:

Equation number	*Dependent variable*	*Explanatory variables*
1.1	I_c	$A_s, I_r, D_w, D_p, D_{z1}, D_{z2}$
1.2	I_d	$A_s, I_r, D_w, D_p, D_{z1}, D_{z2}$
2.1	I_c	$A_s, I_r, D_w, D_{p1}, D_{p2}, D_{z1}, D_{z2}$
2.2	I_d	$A_s, I_r, D_w, D_{p1}, D_{p2}, D_{z1}, D_{z2}$

where :

I_c = conventional index of cropping intensity

I_d = crop-duration index of cropping intensity

D_w = tubewell-used dummy = 1 for tubewell use; 0 for canal use

D_p = tactor-use dummy = 1 for tractor use; 0 for bullock use

D_{p1} = tractor-ownership dummy = 1 when tractor is owned; 0 otherwise

D_{p2} = tractor-hire dummy = 1 when tractor is hired; 0 otherwise

D_{z1} = zone I dummy = 1 for zone I; 0 for other zones

D_{z2} = zone II dummy = 1 for zone II; 0 for other zones

A_s = farm size (net sown area) in hectares

$$I_r = \frac{\text{Gross irrigated area}}{\text{Gross cropped area}} \times 100$$

The equations presented have been computed by the method of ordinary least squares, and specified in the linear form since this gives not merely the direction of the cropping intensity effects but also their absolute magnitudes which are of interest here. (Log-linear specifications were also attempted, but the results were similar to those obtained from the linear specifications. Hence only the latter have been presented here.)

Equations 1.1 and 1.2 help to estimate the effects of tractor and tubewell use in comparison with the use of bullocks and canal respectively. In equations 2.1 and 2.2 the attempt is to disaggregate the overall tractor effect by the criteria of ownership and hire. We can thus compare exclusively bullock farms, tractor-hiring farms and tractor-owning farms. Dummy variable have been specified to separate the effect of the three agro-climatic zones, as well as to mark the differences between tractor-using (owned and hired) and bullock-using farms, and between tubewell-irrigated and canal-irrigated farms.

To identify the tractor and tubewell effects, there were two main reasons for using discrete (dummy) rather than continuous variables as, for instance, taking the percentage of farm area ploughed with tractors and the percentage of farm area irrigated with tubewells: (a) the effect of tractor or tubewell availability (through ownership or hire) on the decision of whether or not to grow an extra new crop or to plant more area with existing crops, is likely to be once-for-all during the year. We would thus expect the impact on cropping intensity to be in the nature of a jump. The more intensive use of a tractor or a tubewell may effect the output produced on the area planted, but it is unlikely to make much difference to the amount of area planted; and (b) in practice, it is not always possible to specify what percentage of farm area is operated with any single technique, because sometimes the same plot of land is given a few ploughings with a tractor and a few with a bullock; and similarly given a few irrigations with a tubewell and a few with a canal.

To compute the crop-duration index, the formula presented earlier has been used. For each farmer, the area under each crop (identified by variety) cultivated by him has been multiplied by the average time-duration of that crop, thus obtaining the total hectare-months for which his land is occupied.

IV. RESULTS AND INTERPRETATION

Tabulation Results

Table 4.1 gives the number and average size of farms using different combinations of ploughing and irrigation techniques; it also indicates, in the case of tractors, whether they are owned or hired. We note that 54.8 per cent of the farmers use tractors. Of these 40.8 per cent own the machine and 59.2 per cent rely exclusively on hired tractors. If tractor ownership alone had been taken as the criterion for separating bullock and tractor farms, these tractor-hiring farms would have been misclassified as bullock farms. The majority—90.8 per cent—of the tractor-using farms are irrigated by tubewells. Among these, those owning tractors are almost all tubewell irrigated. Of the bullock-ploughed farms, 29.0 per cent are exclusively canal irrigated, and the remaining 71.0 per cent are tubewell irrigated. The average farm size for the sample comes to approximately eight hectares. Tractor farms, particularly tractor-owning ones, not unexpectedly, are larger than bullock farms. The latter have an average size of 6.06 hectares compared with the 9.54 hectares of all tractor-using farms. Among tractor-users, owners again are larger (12.85 hectares) than hirers (7.27 hectares).

Table 4.2 gives a cross-tabulation of the mean cropping intensities by combinations of ploughing and irrigation techniques. All the farms in the sample are cultivated in both the *rabi* and the *kharif* seasons. For all farms taken together, the average cropping intensity by the conventional index comes to 168.4 per cent. The mean value of the crop-duration index comes to 69.5, that is, the average utilisation of land over the year is approximately 70 per cent by I_d. Expressed in hectare-months, this would mean that out of the maximum of 12 months for which a hectare could be occupied, it is occupied on average for 8.3 months. The coefficients of variation for both 1_c and 1_d are low, being 0.18 and 0.16 respectively.

TABLE 4.1 : Number of farms and average farm size by farm technique categories

Farm Technique Categories	*No. of Farms*	*Average Farm Size (in Hectares)*
1. All Bullock Farms	107	6.06
(i) Bullock Canal	31	7.33
(ii) Bullock-Tubewell	76	5.54
2. All Tractor Farms	130	9.54
(i) Tractor owners	53	12.85
(a) Tractor owned-Canal	2	15.91
(b) Tractor owned Tubewell	51	12.73
(ii) Tractor hirers	77	7.27
(a) Tractor owned-Canal	10	8.68
(a) Tractor owned-Canal	67	7.06
3. All Farms	237	7.97

TABLE 4.2 : Cropping intensity by farm technique categories (Mean Values)

Farm Technique Categories	*Conventional Index*			*Crop-duration Index*		
	Canal	*Tubewell*	*All*	*Canal*	*Tubewell*	*All*
Bullock	127.8	175.3	161.5 (0.20)	62.6	70.2	68.0 (0.14)
Tractor	134.2	178.1	174.0 (0.15)	65.6	71.4	70.8 (0.16)
Tractor owned	136.2	177.7	176.2 (0.13)	72.8	70.8	70.9 (0.17)
Tractor hired	133.8	178.3	172.6 (0.17)	64.1	71.8	70.8 (0.16)
All	129.6 (0.13)	177.0 (0.14)	168.4 (0.18)	63.4 (0.13)	70.9 (0.16)	69.5 (0.16)

Notes : Figure in brackets give the coefficients of variation of the respective values. For number of farms in each technique category see Table 4.1.

On comparing farms of different technique categories we find that by both indices, bullock-canal farms have the lowest cropping intensity, and tractor-tubewell farms have the highest. In between come tractor-canal farms and bullock-tubewell farms in ascending order. However, the difference between bullock-tubewell and tractor-tubewell farms is small. The additional effect of farm size emerges from the regression equations.

Regression Results

The regression results (see Tables 4.3a and 4.3b) help to bring out the effect of the specified varibles as below :

TABLE 4.3a : Cropping Intensity Regression Results (Linear)

Equation number	1.1	1.2	2.1	2.2
Dependent variable	I_c	I_d	I_c	I_d
Number of observations	237	237	237	237
$\overline{R}^2$	0.56	0.32	0.56	0.32
Constant	99.97	47.08	102.79	48.26
Explanatory variables	*Regression Coefficients*			
Farm Size (A_s)	–1.29**	–0.57**	–1.52**	–0.66**
	(4.93)	(4.76)	(5.26)	(5.04)
Per cent Area Irrigated (I_r)	0.39**	0.22**	0.38**	0.22**
	(4.73)	(5.82)	(4.65)	(5.75)
Tubewell-use dummy (D_w)	34.51**	13.30**	34.07**	13.11**
	(6.73)	(5.69)	(6.67)	(5.62)
Tractor-use dummy (D_p)	6.77**	2.46	—	—
	(2.35)	(1.88)		
Tractor-ownership dummy (D_{p1})	—	—	12.06**	4.68**
			(2.97)	(2.53)
Tractor-hire dummy (D_{p2})	—	—	4.86	1.67
			(1.60)	(1.20)
Zone-I dummy (D_{z1})	23.40**	–4.49*	22.31**	–4.95**
	(4.76)	(2.00)	(4.53)	(2.20)
Zone-II dummy (D_{z2})	12.19*	–6.85**	11.76*	–7.03**
	(2.47)	(3.05)	(2.39)	(3.14)

Notes : Figures in brackets give the 't' values of the respective coefficients.
**Denotes significance at the 1% level.
* Denotes significance at the 5% level.
The rest are insignificant at the 5% level.
Significance has been tested in accordance with the two-tailed test.

(i) *The use of tractors and tubewells :* Equations 1.1 and 1.2 confirm that farms using tubewells have a significantly higher mean cropping intensity than those relying on canal water alone: the difference comes to 34.51 percentage points by the conventional index and 13.30 by the crop-duration index. Again, by the conventional index, tractor-using farms are associated with a significantly higher cropping intensity than bullock farms, the difference coming to 6.77 percentage points. By the

TABLE 4.3b : Differences in Regression Coefficients of Dummy-Pairs (Linear Results)

Equations number	*Dummies compared*	*Difference in the coefficients of the dummies compared**	*'t' values of the differences*
1.1	D_{z1}, D_{z2}	11.21**	3.54
1.2	D_{z1}, D_{z2}	2.36	1.63
2.1	D_{p1}, D_{p2}	7.20**	2.83
2.1	D_{z1}, D_{z2}	10.55**	3.32
2.2	D_{p1}, D_{p2}	3.01	1.68
2.2	D_{z1}, D_{z2}	2.08	1.44

Notes : * The coefficient of the dummy mentioned second has been subtracted from the coefficient of the dummy mentioned first.

** Denotes significant at the 1% level.

The rest are insignificant at the 5% level.

Significance is in accordance with the two-tailed test.

crop-duration index, however, the differences between bullock and tractor farms are not significant;

(ii) *Ownership of a tractor :* From equations 2.1 and 2.2, we obtain the effect not merely of the use of tractors but also of their being owned or hired. Tractor-owning farms are found to have a significantly higher cropping intensity, by both indices, than bullock farms: the difference comes to 12.06 as measured by I_c and 4.68 as measured by I_d. Tractor hirers too have a higher cropping intensity than bullock farms but here the differences are not statistically significant by either index. Relative to tractor hirers, tractor owners have a significant advantage by the conventional index. Also, the tubewell-irrigated farms are again (among this second set of equations) found to have a significantly higher mean cropping intensity than canal-irrigated farms, by both indices;

(iii) *Other factors : per cent area irrigated, farm size, agro-climatic zones :* The index of irrigation is positively related to cropping intensity in all four equations. However, the coefficients of the irrigation variable, though significant, are small. This would be because, in the majority of farms, over 85 per cent of the gross cropped area is already irrigated and subsequent increases at the margin would thus lead to only small additions to cropping intensity.

Farm size is consistently significant and inversely related to cropping intensity in all the equations, as hypothesised.

Finally, both the zonal dummies emerge as significant variables in all the equation. It is interesting to note that the results for the conventional index have a sign opposite to those for the crop-duration index. In equations pertaining to the conventional index, Zones I and II have a significantly higher cropping intensity than Zone III; Zone I has the highest value, followed by Zones II and III. In equations pertaining to the crop-duration index, however Zone III emerges as the one with the highest cropping intensity. This difference in the results obtained with the two indices may be attributed to the difference in the time-duration of crops grown in the three zones in the kharif season. The main rabi crop, namely HYV wheat, is the same in all the zones, but the main *kharif* crop differs: in Zone III it is American cotton, which takes seven to eight months to mature, while in Zones I and II the main *kharif* crops are maize, paddy and groundnut which have an average duration of four to four and a half months. Hence Zone III, which has the lowest cropping intensity by the conventional index, has the highest by the crop-duration index.

Using the regression results, I also calculated the additional land that would be made available to the farmer over the crop year, due to the observed increase in cropping intensity through tractor and tubewell use. For an eight-hectare farm (which, as noted, is the average farm size in the sample), the land made available through increases in cropping intensity as measured by I_c came to 0.96 hectares for the owned tractor relative to bullocks, and to 2.72 hectares for the tubewell relative to canal.[6]

V. SUMMARY AND CONCLUDING COMMENTS

This paper set out to identify the factors which could explain the observed differences between farms in their cropping intensities. In particular, I wanted to quantify the effect of tractors and tubewells, and to see whether the ownership of a tractor gave the user any special advantage over one who merely hired the machine. To take account of differences in the duration of different crops, I specified, in addition to the conventional index, the crop-duration index of cropping intensity which could also give an idea of the degree to which a farmer under-utilised the land over the crop-year.

Both tractors (in particular owned tractors) and tubewells were found to be associated with a higher cropping intensity, as measured by each of the indices, in comparison with bullocks and canals respectively. However, the advantage of tubewells over canals was much greater

(34.1 percentage points as measured by I_c) than that of tractors over bullocks (which came to 12.1 percentage points). That irrigation, in particular, tubewell irrigation, is a more important factor in increasing cropping intensity, than the use of tractors in lieu of bullocks, has been emphasised in other studies as well [see Billings and Singh on the Indian Punjab; Kaneda, McInerney and Donaldson for Pakistan].

Likewise, the association of zones with differences in cropping intensity noted here, agrees with the findings of Vashishtha who also took this factor into account (although, unlike here, he did not spell out why one zone may have an advantage over another). In this zone-wise comparison, the alternative crop-duration index helped to highlight the importance of taking account of cropping pattern differences, especially where farmers grow crops with substantially differing maturing periods.

The negative relationship that we observed between farm size and cropping intensity, for all farms taken together, again is in keeping with the results of others such as Vashishtha, Motilal, Rao, Grewal and Kahlon, and McInerney and Donaldson. Also, it is noteworthy, although the detailed results are not presented in this paper, that in my sample even when farms were separated into bullock, tractor-owning and tractor-hiring categories, the inverse relationship held for all three categories. This would caution against the contention of some scholars that tractors help in *preventing* a fall in cropping intensity on large farms [see, for example, Vashishtha, 1975:41]. The most that could be said is that given the observed advantage of tractors over bullocks, cropping intensity would not fall to the same *level* on large tractor farms as on large bullock ones.

In this context, it is also necessary to view with caution the popular notion that tractors are likely to release large amounts of fodder land for crop cultivation. Although no precise estimates on this could be made from my data, the differences between bullock and tractor (owned or hired) farms in the average percentage of gross cropped area (GCA) devoted to fodder, was not found to be substantial: bullock farms devoted 16 per cent, tractor-hiring farms 15.6 per cent, and tractor owning farms 11.6 per cent of GCA to fodder.

On a *priori* grounds too, we would not expect much saving from fodder land with tractors because : (a) only a part of the bullock feed usually comes from fodder grown on the farm; much of it often consists of wheat and rice straw which are by-products (Bose and Clark, 1969).

Hence, even on bullock farms the amount of land needed for fodder would be small; (b) on tractors-hiring farms, bullocks are often retained to guard against the risk of tractors not being available for hire when needed; and even on tractor-owning farms, bullocks are often kept for some agriculatural operations, and to counter the risk of a tractor breakdown especially during periods of peak requirements; (c) farms may 'fit in' a short duration fodder variety between the main crop seasons, in effect using land which would otherwise lie idle for that period; (d) fodder may be grown for sale, depending on the profitability of fodder relative to other crops.

On individual farms, therefore, tractors need not lead to a diversion of land under fodder to other crops; and at the macro-level, say, for Punjab as a whole, whether or not tractorisation significantly reduces the area under fodder would depend on how substantial a reduction in bullock requirements and hence in the bullock population it helps to bring about.

On the whole, the cropping intensity advantages of tractorisation in the Punjab appear to have been overstated.

Note : Reproduced from *The Journal of Development Studies*, July 1984 with kind permission of the publisher.

NOTES

1. The studies mentioned here are for purpose of illustration. For a more extensive coverage of research on the cropping intensity effect of tractorisation, see Binswanger's review (1978). Binswanger does not, however, highlight the particular shortcomings of this literature that are emphasised here.
2. Gross cropped area is the total area under all crops grown over the crop-year. Net sown area is that part of the farm area that is cultivated at least once over the crop-year; it is given by: owned area plus net leased-in area minus area unclutivated throughout the year minus area under orchards or gardens.
3. The main sample had 240 farms; three were dropped due to incomplete data.
4. The original classification of zones adopted by the data-collecting unit did not strictly correspond to its stated criteria. Hence, for the present analysis, some *tehsils* were reclassified so that those falling within a zone were broadly homogeneous in three terms.
5. The principal agricultural operation for which tractors were used by the sample farmers was ploughing (concluding subsidiary functions associated with seed-bed preparation). Hence the separation of farms into tractor-using and bullock (non-tractor) using is based on the use or non-use of tractor in

this operation. Some farms, however, sometimes did use tractors for additional operations as well.

6. This is given by multiplying the average farm size (in this case approximately eight hectares) with the regression coefficients obtained in equation 2.1 and dividing by 100.

REFERENCES

Agarwal, Bina, (1977), Mechanization in Farm Operations—Choice and Their Implications. A Study Based on Punjab, Ph.D. dissertation in Economics, University of Delhi; to be published as *Mechanization in Indian Agriculture: An Analytical Study Based on the Punjab,* Delhi : Allied Publishers.

Agarwal, Bina, (1980), 'Tractorisation, Productivity and Employment : A Reassessment', *The Journal of Development Studies,* Vol. 16 No. 3.

Billings, Martin and Arjan Singh, 1969, 'Labour and the Green Revolution — The Experience of Punjab', *Economic and Political Weekly,* Vol. 4 No. 52

Binswanger, Hans P., (1978), *The Economics of Tractors in South Asia : An Analytical Review,* New York : Agricultural Development Council; Hyderabad : International Crops Research Institute for the Semi-Arid Tropics.

Bose, Swadesh R. and Edwin H. Clark II, (1969), 'Some Basic Considerations on Agricultural Mechanisation in West Pakistan', *Pakistan Develpment Review,* Vol 9, No. 3.

Chopra, Kusum, (1972), *Tractorisation and its Impact on the Agrarian Economy of Punjab,* New Delhi: USAID, February.

Grewal, S.S. and A.S. Kahlon, (1972), 'Impact of Mechanisation on Farm Employment in Punjab', *Indian Journal of Agricultural Economics,* Vol. 27 No. 4.

Johl, S.S., (1970), 'Mechanisation, Labour-Use and Productivity in Indian Agriculture, Economics and Sociology', Occasional Paper No. 23, Ohio State University, Columbus, Ohio.

Kaneda, Hiromitsu, (1969), 'Economic Implications of the "Green Revolution" and the Strategy of Agricultural Development in West Pakistan', *Pakistan Development Review,* Vol. 9, No. 2.

Mclnerney, John P. and Graham F. Donaldson, (1975), 'The Consequences of Farm Tractors in Pakistan', IBRD Bank Working Paper No. 210.

Motilal, G., (1973), 'Economics of Tractor Utilisation', *Indian Journal of Agricultural Economics,* Vol. 28. No. 1.

NCAER, (1973), *Impact of Mechanisation in Agriculture on Employment,* New Delhi: National Council for Applied Economics Research.

Rao, C. H. Hanumantha, (1975), *Technological Change and Distribution of Gains in Indian Agriculture,* New Delhi: Macmillan.

Vashishtha, Prem, (1975), 'Issue in Technological Adaptation and Agricultural Development: An Analysis of Production Functions on Punjab Farms', Ph.D. dissertation in Economics, Delhi University.

5

SUBSTITUTION AND COMPLEMENTARITY BETWEEN INPUTS IN PADDY CULTIVATION

KANCHAN CHOPRA

I. THE ISSUE

The issue of substitution possibilities between inputs in production has been viewed from many angles in the literature. The early attempts were couched in terms of the two-factor capital-labour substitution problem. This was viewed essentially in the light of economy wide resource-endowments and the appropriateness or otherwise of a technology was judged by looking at the nature of the technology relative to these resource endowments. It was in this context that discussions in the fifties tackled the problem of factor substitution.

Following from this, each technology or technique of production came to be evaluated in terms of its effect of some over all objective of the economy. For a long time, the employment *vis-a-vis* the production aspects of techniques were the focus of discussion. In the field of agriculture, mechanisation in the form of introduction of tractors, tubewell water and threshers led to the use of factor inputs in different proportions from what had been used hitherto. The leading preoccupation of studies on mechanisation seems to have been the issue of what impact they had on employment and, at most, cropping intensity and output. It was also suggested that the level of aggregation may have a significant impact on the conclusion. In the field of industry too, arguments on factor-intensity in small scale versus large-scale industry seem to have been dwelling on the employment and production aspects of different scales and techniques.

These explanations were looking at factor intensity as a means of achieving economy-level planning objectives. Of late, the appropriateness of viewing substitution from a two-factor view point has been questioned in the literature. Natural resource based raw materials and energy are two other significant inputs and it is not always analytically correct to subsume one or the other under capital. This can only be done under limiting assumptions relating to the nature of substitution between labour, capital and raw materials, more specifically natural resource-based raw materials. In a more general framework, it becomes necessary to extend the analysis to a multi-input situation. The operational significance of this increases in view of the rising prices and the exhaustible nature of some sources of energy. Two issues then arise : (1) the significance of the substitution-complementarity relationships from the viewpoint of the availability of resources in the economy, (2) The methodology to be used to extend the analysis to a multi-input framework. The first places the problem in its analytical perspective whereas the second is a methodological issue.

Viewing a technology in a multi-input perspective has a direct link with the problem of exhaustible natural resources. Whether or not exhaustible resources shall present an ultimate constraint to growth depends on whether labour, capital and renewable resources can be substituted for them in production.[1] This has been illustrated, among others, by Dasgupta and Heal.[2] Though their models are aggregate longrun models, the information necessary to use the logic they present has to originate at the level of empirical studies based on micro data. Our exercise attempts to present one such study with reference to paddy cultivation in Indian agriculture. Indian agriculture in the early eighties presents a scene of wide variations in technology. The new HYV technology, in particular, uses a variety of inputs, (e.g. fertilisers, controlled supply of water, hired labour) that were not a part of traditional technology in agriculture. The possibility of substitution or otherwise between these inputs in the new technology shall throw light on the longrun constraints to growth in view of the relatively limited supplies of some of these inputs. It is in this context that we are examining the new agricultural technology represented by the adoption of HYV varieties in paddy in the north-west region. Hence, an important issue is—what are the possibilities of substitution between water, land, fertilisers, labour and capital in this technology. The nature of the technology would also have an impact on the implications of agricultural policy relating to pricing of inputs. Subsidies or price rises could only be expected to have an impact on input use if the technology has some degree of flexibility.

We are studying paddy cultivation in Haryana. The varieties grown are mainly Jaya, IR8, and Basmati. All these yield a high value of output per hactare.[3] Inputs used are canal irrigation, chemical fertilisers, tractor hours (owned or hired) and owned and hired labour. There is very little use of organic manure. It is in this context that we examine substitution-complementarity relationships between inputs. A priori we have evidence to show variation in input use per hectare. This variation is observed along with a variation in input prices.[4] This leads us to believe that the problem can be viewed in a cost function framework in which input prices and output are the independent variables.

II. THE METHODOLOGY

The methodology followed in this paper (i) specifies a cost function with input prices and scale of output as independent variables and (ii) derives the parameters of the production technology, i.e., the elasticities of substitution between inputs and the elasticities of factor demand, from the parameters of the cost share equations corresponding to the cost function. This approach has been used to estimate elasticities of substitution in application to both agriculture and industry.[5] A cost function approach has advantages over the production function approach with regard to both conceptualisation and estimation procedure. Binswanger has outlined some of these advantages. It views the cultivator as a cost minimiser treating input prices as exogenous variables. In the north-west region of India to which our study relates, it might be justifiable to view the cultivator in the above light; agriculture is a commercialised venture and cultivators can reasonably be expected to be sensitive to input prices. Also, there is variation in input prices and in input use between farm. Input prices vary and each cultivator believes that these variations are caused by factors beyond his control, particularly in the short-run. He tends to minimise costs with given input prices which show cross sectional variations.

The notion of cross-section variation in prices needs some explanation. It is probably less plausible than the notion of variability over time. It arises out of imperfect access to inputs or to knowledge. Input data collected from farm households on the basis of a cost accounting approach reveals variation in both, prices and level of use. These prices, basically perceived by the cultivator, reflect in turn an array of factors, such as, for instance, differential access to retail outlets and subsequent costs of transport. Price, therefore, enters as a proxy for a number of other variables into the farmers decision making process which is a very complex one.

Our primary interest in this study is in estimating substitution between inputs in multi-input framework. In deciding on the form of the cost function, this had to be the primary consideration. All constant elasticity of substitution forms were, therefore ruled out. The transcendental logarithmic cost function which allows for variable elasticities of substitution seemed most appropriate. In recent empirical work, this has been widely used. In a recent study on wheat in the Punjab, for example, it has been suggested that the translog provided a better aproximation to the structure of production than the Cobb-Douglas.[6]

The translog cost function helps us to estimate the substitution elasticities for the production function corresponding to it. It does not, however, throw light on the exact form of the underlying production function. We are able to comment on the structure of production only in so far as it relates to the elasticities of substitution and of input demand. This limitation does not hamper our analysis since the latter two form the main focus of our analysis.

Another methodological issue that has some bearing on the results is the number of inputs. We shall postulate a cost function with six inputs, land, fertiliser, water, labour and conventional and modern capital. This seems to us to be sufficiently disaggregated for paddy cultivation in Haryana. These are the important inputs and there are different levels of possible exhaustion of supply linked to them. Land in probably the one in the most inelastic supply. The known reserves of surface and groudwater in the north-west region have also been tapped and it does not seem that any augmentation in its supplies can be hoped for.[7] It is, therefore, important to see to what extent, the other factors can be substituted for them. In this paper, we are examining this issue in the context of paddy cultivation.

The ultimate criterion for determining whether some inputs can be aggregated with others lies in the separability tests discussed extensively in the literature.[8]

We shall, carry out a few separability tests to determine whether the level of disaggregation we use is justified by the nature of the substitution relationships between the inputs we consider.

The cost function is specified as

$$\log C = V_o + V_q + \log Q + \sum_{i=1}^{n} V_i \log P_i + \frac{1}{2}\sum_{i=1}^{n}\sum_{j=1}^{n} b_{ij} \log P_i \log P_j$$

$$+ \sum_{i=1}^{n} b_{iq} \log P_i \log Q \text{ ,} \qquad \text{.......(1)}$$

where C = total cost; P_i , P_j = prices of inputs i, j; Q = output of unit, b_{ij}; and V_o, V_q, V_i, b_{ij}, b_{iq} are the parameters of the funciton. The parameters of the cost function, which describe the production technology, can be estimated from the cost share equations. Using Shepherd's Lemma, the cost share equations can be derived as[9]

$$\frac{\partial \log (C)}{\partial \log (P_i)} = C_i = V_i + \sum_{i=1}^{n} b_{ij} \log P_j + b_{iq} \log Q \qquad \text{.......(2)}$$

$$i = 1, 2, \ldots . \ n,$$

where C_i = share of the ith input in total cost. The parameters of the cost share questions are estimated subject to the following restrictions arising out of the assumption of linear homogeneity and the requirement that the cost shares add to 1. The restriction are :

$$\sum_{i=1}^{n} V_i = 1 \text{ , } \sum_{i=1}^{n} b_{ij} = \sum_{j=1}^{n} b_{ij} = 0 \text{ and } \sum_{i=1}^{n} b_{iq} = 0^{10}. \qquad \text{.......(3)}$$

The parameters of the cost share equations and the cost shares are then used to derive the Allen partial elasticities of substitution (AES) between inputs i and j, h_{ij}[11]

$$h_{ij} = \frac{b_{ij} + C_i C_j}{C_i C_j} \qquad i \neq j \text{ , } j = 1, \ldots n.$$

$$h_{ii} = \frac{b_{ii} + C_i^2 - C_i}{C_i^2} \quad i = 1, \ldots . n. \qquad \text{.......(4)}$$

Further, the price elasticities of demand for inputs (e_{ij}) can also be drived from the h_{ij}'s.

$$e_{ij} = C_j h_{ij} \text{ , } e_{ii} = C_i h_{ii} \text{ .} \qquad \text{.......(5)}$$

Both h_{ij}'s and e_{ij}'s are the crucial parameters in the production technology. In the above framework, h_{ij} can be tested for statistical significance using an approximate standard error obtained by assuming that C_i's are non-stochastic. The parameters found to be significant are sometimes referred to as the 'informative' elasticities of substitution.[12] The conclusion arrived at in the literature seems to suggest that these informative elasticities are not likely to err in identifying the nature of the association between two inputs, though they may err in so far as percise magnitudes are concerned. The AES can then be taken to measure the substitution relationships between inputs with a fair degree of correctness.

To estimate the cost share equations (2) subject to the restrictions (3) we use a standard iterative three stage least squares (I3SLS) procedure. This iteration makes the estimated coefficients and residual covariance matrix converge and the estimates are an improvement over the 3 SLS estimates which may not be invariant to the equation deleted in the estimation.[13]

III. THE DATA BASE

We have used farm level data provided by the Cost of Cultivation surveys of the Ministry of Agriculture. These surveys provide extensive data on quantity of inputs, output and asset structure. The design of the survey is three stage stratified random sampling with the tehsil as the first stage, the cluster as the second and the operational holding as the third stage sampling unit. Each state is divided into a number of homogeneous zones on the basis of cropping pattern, soil types and rainfall and tehsils are selected from each zone. The data is collected on a continuing basis every year since 1971-72.

The regions chosen to represent cost of cultivation for different crops vary from year to year and Punjab was included in the main sample for paddy in 1974-75.[14] Haryana was included in the main sample for paddy for 1977-78. We did not want to use data for 1974-75, since it was a bad agricultural year.[15]

We decided, therefore, to use the Haryana data for 1977-78. This was a normal agricultural year and the technological choices and sociological factors are more or less similar in the two states. Though, Haryana has had somewhat lower rates of agricultural growth than Punjab, the spread of paddy has been under similar conditions and the factors affecting input relationships, we feel, would be broadly similar.

Our sample consists of sixty farms selected from the above survey. These are spread over the two agroclimatic zones into which Haryana falls, forty falling in one zone and the rest in the other. This comprises a part of the total number of farms included in the sample in that year. This limitation to a subsample was made inevitable by (a) non-availability of data relating to some clusters, and (b) the need to pick up farms from both agroclimatic zones to achieve a fair degree of representation for each. The size of the sample finally taken is however, large enough to enable statistically significant results to be arrived at.

The inputs are land, water, fertiliser, human labour, conventional capital and modern capital. These are the important inputs with land constituting on an average, 34.6 per cent of the cost of production, water 10.4 per cent, fertiliser 21.6 per cent, human labour 22.6 per cent and capital 11.3 per cent (conventional capital comprising an average of 2 per cent of the total cost and modern capital 9.3 per cent). We give below a brief account of the construction of the variables relevant for our analysis i.e. the input prices, cost shares, total cost and level of output for each farm.

Rent per Hectare and Cost of Land per Farm

The farm level schedules give an imputed rental for each farm. The rent imputed is equal to 'the rents prevailing in the area.' As leasing-in is almost negligible, this imputed rent is the price of land in our sample. The 'imputed rents' area of course, uniform as between farms falling in the same area. Our sample extends however to different agro-climatic regions and this leads us to expect, *a priori,* variations in land rent. The data too, reveals variations, the rent per hectare varies from Rs. 512 per hectare to Rs. 1158 per hectare. The corresponding cost of land is computed by multiplying rent by the area under paddy in each farm.

Price per Hour of Water and Cost of Water per Farm

Fifty-three farms use tubewell water for paddy, the rest use canal water. Tubewell users fall into two categories; those who buy tubewell water and these who own tubewells. A per hour cost is given for the first category and this, when multiplied by the number of hours of tubewell use in paddy (obtained from detailed schedules of cropwise machinery use data) gives the cost for the tubewell water for the hiring farms. For farms that own tubewells, the cost of an hour of tubewell use is calculated by taking the total of interest, depreciation and operating expenses. The last item includes the cost of diesel, electricity and minor maintenance. Interest is taken as a pure time-based cost whereas depreciation is taken to be a function of both time and use. The cost of the services of tubewells thus obtained is based on the user cost concept.[16] Multiplying this per hour cost by the number of hours of owned tubewell use for paddy gives the cost of water for these farms. For canalwater using farms, the per hour price for canal users is arrived at by taking an average figure for number of hours of water use per hectare under paddy cultivation.

Price of Fertiliser per Farm and Cost of Fertiliser per Farm

For fertiliser, the data available is less satisfactory. Value of fertiliser used and quantity used on paddy per farm are both given. An average price is arrived at by dividing one by the other. There is the danger that this average may reflect variations in proportions of the three components (*N, P* and *K*) rather than variations in price which should be the correct independent variable. To check this, we looked into the proportion of *N, P* and *K* Farmwise. Though the proportions did vary, there was a large number of farms using *N* only and the range of variation in price could be explained by variations relevant for farms that were using *N* only. We could, therefore, conclude that the range of variation in the price was caused by difference in the purchase price of fertiliser. A part of inter-farm variation may, however, have been on account of differences in proportions used of the three components. *A priori,* differences in purchase price of fertiliser could be due to difference in time and agency of purchase. Organic fertiliser was not included since its use was marginal and there was no interfarm variation in its price.

Wage Rate and the Cost of Human Labour

Family labour hours and hired labour hours in paddy production are recorded separately and the payments made to hired labour are given.

A wage rate per hour for hired labour is worked our for each farm and family labour is also valued at this wage rated to yield cost of labour in paddy production. This is done on a farm level, the rates at which labour is hired in by the farm are used to evaluate family labour.

A priori, one would expect to find very little variation in wage rate in the same geographical region. We find, however, that the small farmers are systematically paying higher wage rates for hired labour. This reflects in part their lower bargaining power in the market for labour, arising, may be, out of lower levels of demand.

Price of the Services of Capital, and the Cost of Capital

Capital includes conventional equipment (i.e. plough, sohaga etc.) draught animals and tractors. There is no hiring of animals and conventional equipment. Also, thirteen farms own tractors. For all these owned assets, the cost of services of capital is estimated as in the case of owned tubewells.[17] It is the sum of interest and depreciation cost, operating costs, minor repair and maintenance. For farms hiring tractors, cost of hiring per hour is given. Information on number of hours of use in paddy production for tractors and drought animals is given. Conventional equipment is used in conjuction with drought animals in most operations. So for conventional capital we take number of hours of animal power use as the relevant measure of use in paddy production.

Cost Shares are defined as cost of a particular input as a proportion of total cost. *Quantity of Output, Q* is the production of paddy per farm, measured in quintals. By-products are included in the value of output after being converted into equivalent quintals of paddy.

IV. ANALYSIS OF RESULTS

We obtain estimates of the parameters of the six-input translog cost function by applying iterative 3SLS to the system of cost-share equations. Table 5.1 gives the parameter estimates. A large number of the parameter estimates are significantly different from zero. This means that the coefficients are significant and the Cobb-Douglas, (which is a special case of the trans-log when the coefficients of the second order terms are zero) would not be a correct approximation of the cost function. As the Cobb-Douglas is a self-dual, the production technology could also then not be correctly characterised by it.

The AES derived from the parameter estimates are given in Table 5.2 of the text. A number of estimates of elasticity of substitution are significant at the 1 per cent or 5 per cent levels. There is a considerable degree of complementarity between land, labour, and, to some extent, conventional capital. Land can, however, be substituted for by water and modern capital, both of which are in turn substitutable by fertiliser. In other words, availability of irrigation water and mechanical equipment could encourage more intensive use of land.[18] The use of the latter (tractors in our sample) would be at the expense of use of human labour, as we find that a significant substitutability exists between modern capital and labour.

TABLE 5.1 : Estimates of Parameters of Cost Share Equations (Haryana Paddy, 1977-78, Six Inputs)

Cost Share equation for	*Price of Land*	*Price of Water*	*Price of Fertilizer*	*Price of Labour*	*Price of C. Capital*	*Price of M. Capital*	*Level of output*
1. Land	0.37**	.025	–0.14**	-0.21**	–0.04	–0.008	0.18
	(9.25)	(1.84)	(–3.36)	(–8.23)	(–1.9)	(–2.03)	(1.05)
2. Water	—	–0.04**	–.066**	–.048**	–0.006	.0087*	–.009
		(–2.48)	(4.50)	(3.44)	(–.55)	(2.91)	(–.42)
3. Fertilizers	—	—	.09	–0.003	–0.005	–0.002	0.056**
			(1.7)	(–0.43)	(–0.29)	(–0.56)	(3.08)
4. Labour	—	—	—	0.28**	–0.005	+0.002	–0.04*
				(8.63)	(–0.29)	(–0.396)	(–2.5)
5. Conventional Capital	—	—	—	—	0.154	–0.006	–0.018
					(6.3)	(–0.18)	(–1.1)
6. Modern Capital	—	—	—	—	—	0.005**	–0.002
						(4.49)	(–0.65)

Notes : 1. As the matrix is symetric, the entire matrix is not reproduced.
2. *stands for significant at 5% and
* *stands for significant at 1% in all tables.
3. weighted R^2 for the system is .45.

TABLE 5.2 : Allen Elasticities of Substitution (Haryana Paddy; 1977-78, Six Inputs)

	Land	*Water*	*Fertilizer*	*H. Labour*	*C. Capital*
Water	1.714**	—	—	—	—
	(3.99)				
Fertilizer	–0.818	4.00**	—	—	—
	(–1.58)	(6.29)			
Human Labour	–1.62**	–1.087	0.745	—	—
	(–5.11)	(–1.79)	(1.22)		
Conventional Capital	+.43	–2.0	–0.204	–0.52	—
Modern Capital	0.75**	1.96**	0.86**	0.918**	1.33
	(5.35)	(5.93)	(3.63)	(4.52)	(0.647)

Notes : 1. A positive (negative) value of elasticity implies substitutability (complementarity).
2. *Denotes significant at 5% level; and
**Denotes significant at 1% level.

A degree of substitutability between land and machinery and between labour and machinery has been reported by Binswanger, for American Agriculture and by Kako for paddy cultivation in Japan.[19] Though both studies use the same methodology as ours, Kako's study on paddy cultivation is at a more disaggregated level and therefore, comparable. His estimates are numerically lower than ours and are not tested for significance. In both the above mentioned studies, there is a category 'other inputs'. Whenever, there is a residual category of the above kind, it is not easy to interpret the results, particularly when large values of elasticity appear under that category. Sidhu and Baanante (1980) have estimated substitution elasticities for Punjab wheat for 1970-71 and find substitutability between capital equipment and other inputs. Our results can also be viewed in the light of the literature of the effects of tractorisation on cropping intensity, cropping pattern yield and labour use per hectare. By and large, some increases in cropping intensity have been reported by most studies on tractorisation in the Punjab-Haryana region. Kahlon (1976), Chopra (1974), Sharma (1974), the Govt. of Punjab Study, McIkerney and Donaldson (1975) and Agarwal (1983) all report different degrees of increase in cropping intensity while Rao (1975) and Vashishtha (1975) report no impact on cropping intensity. The substitution between land and tractors in our study seems to suggest that in the face of inelastic supplies of land, tractors could help to increase cropping intensity.[20]

On labour use per hectare, the evidence available in the literature seems to suggest that the use of a tractor per se is associated neither with an increase, nor with a decrease. It would depend on the operation for which it is being used. In our sample, tractors are being used only for ploughing and they substitute for human labour. Raj Krishna decomposes the change in labour input per hectare into a number of effects and finds that tractor ploughing leads to a decrease of 5.26 man hours per hectare. Agarwal also finds that tractors used for ploughing substitute for human labour.[21] Our methodology can claim to establish these results more convincingly as (a) it controls the cropping pattern and yield effects more effectively than can be done by studies that look at differences in input-use on bullock farms as against tractor-farms, (b) it looks at the relationship between tractors and human labour in a multi-input framework.

With respect to fertilizer, there are two significant elasticity relationships.

Both modern capital and water seems to be substitutable for it. We take up first the substitutability between fertiliser and water. This seems counter-intuitive, contradicting the somewhat established complementarity between water and fertiliser. We shall postulate the hypothesis that this complementarity is only a sufficiency relationship implying that, in order to use chemical fertiliser effectively, a certain minimum supply of water is essential. The role of irrigation water as an input changes with the level of water use and the nature of the technology.[22]

Agronomic explanations have also been given for this. At higher levels of water use, the growth of blue green algae may take place and this may provide a kind of organic fertiliser that is a substitute for chemical fertiliser. Another possibility is that the mineral content of the water (which becomes significant when it is used in large quantities) may react on the soil and increase the organic fertiliser content, thereby decreasing the need for inorganic fertiliser.

All the above explanations seem to suggest a two-phase relationship between water and fertiliser. The substitutability relationship comes into its own only after the level of water use reaches a certain threshold. This hunch is corroborated by a few trial runs with wheat data. Here the levels of water use are lower and it does not appear as a significant substitute for fertiliser.

The substitutability between fertiliser and modern capital could imply that where fertiliser is in use, the significance of the ploughing operation is diminished so that the use of tractors can be decreased. Earlier studies that have been sufficiently disaggregated have also arrived at similar conclusions.[23]

Two important elasticity estimates that do not conclusively establish substitutability or complementarity are the ones that show the relationship between human labour and fertiliser and between human labour and water. The first corroborates the findings of earlier studies on the lack of any significant association between fertiliser use and labour intensity. As regards the use of water and labour intensity, the evidence is that tubewells lead to increased use of human labour. But as most of the farms in our sample are using tubewell water for paddy cultivation, the non-users are not sufficient large in number for us to be able to capture that effect.

The coefficients of the variable Q enable us to examine the size-input use relationship. The statistically significant relationships are those of human labour and fertilizer. Bigger farms use less labour and more fertiliser per hectare. The use of tractor is not significantly different across size class of farms, possibly because of the large-scale presence of a market for tractor services. Note, however, that the cost of an hour of tractor use is higher for farms hiring their services, mostly small farms. There may also be differences in the timeliness and ease of availability. The use of water is, predictably, not related to size. Intensity of land use is not related to size either. This is in agreement with recent contentions that with the introduction of high yielding varieties, fertilisers and tractors, the large farmers have caught up with the smaller ones in the intensity of land use and the much discussed inverse relationship between farm size and intensity of land use no longer holds true.

Some Separability Tests and their Implication

Separability tests are designed to help in deciding on the level of aggregation of inputs. If the marginal rate of substitution between any two inputs is affected by changes in the level of the third input, then the third input must be treated as a separate factor in production.[24] It has been suggested that progressive aggregation of inputs will decrease the information that is obtained from the exercise. Separability tests help to decide whether it also influences the nature of substitution and complementarity between other inputs.

The separability restrictions are equivalent to certain equality restrictions on the AES which in turn, imply restrictions on the coefficients of the cost share equations. An *F*-ratio is computed to test the null hypothesis of equality of these coefficients to each other and to zero. If the computed *F*-ratio is higher than tabulated *F*-ratio, the null hypothesis of equality between coefficients and consequently elasticites is rejected.[25] We have conducted three separability tests. The results are tabulated in Table 5.3.

TABLE 5.3 : Tests of Separability

Test Identification	*Calculated F-Value*	*Tabulated F-Value*	*Null Hypothesis*
1. Water from land and fertiliser	16.34	3.19	Rejected
2. Water from modern capital and fertiliser	14.81	3,19	Rejected
3. Modern capital from human labour and traditional capital	0.12	3.19	Accepted

In two cases, both relating to water, the null hypothesis that the substitution elasticities between other inputs are independent of the level of use of water is rejected. Water should then be treated as a separate input. It cannot be hypothesised that water can be subsumed under irrigated land or fertiliser. The contention that there is double-counting of cost involved in taking water as a separate input when the rent is defined as rent on irrigated land is not, therefore upheld by our statistical test. Intuitively, one would say that there are differences in the availability and cost of water on tubewell owning farms, tubewell hiring farms and canal-irrigated farms. These differences justify the treatment of water as a separate input as they are not captured fully in the increased rent of land. Similarly, water needs to be treated as a separate input from fertilizer and capital.

Modern capital being treated as a separate input does not seem to influence the possibilities of substitution between other inputs. The disaggregation helps us to get additional information, of course. If, however, we were to aggregate all capital into one input, it would not influence the direction of the other results. We find that the results are invariant with respect to the other inputs. The substitutability between water and capital and fertiliser and capital remains in the five input case too. The interesting insights on the relationship between modern-capital (essentially tractors) and other inputs are lost. The number of non-informative estimates also increase.[26] We would, therefore, opt for our earlier level of disaggregation.

Price Elasticities of Input Demand

Our main focus in this paper has been on the estimation of substitution and complementarity relationships between inputs. An interesting extension is the calculation of own an cross price elasticities of input demand. This can be done easily using equation (5). We give in Table 5.4 price elasticities of demand with respect to the prices of irrigation-water, fertiliser and modern capital, the three main purchased inputs.

Most price elasticities of input demand have a value of less than one. The only exception is the own price elasticity of irrigation water. Water seems to have relatively higher values for cross elasticities of demand with respect to land and fertiliser as well. An increase in the price of water increases the demand for land. This could be interpreted as a choice between the extensive and intensive margins of cultivation. The own price elasticity of fertiliser on the other hand, is rather low. This would seem to indicate that changes in fertiliser price do not effect demand very significantly. Modern capital is a substitute for other inputs, in particular for human labour, but the elasticity has a value of 0.08 only. Demand for labour, therefore, is not likely to be affected greatly by changes in the price of modern machinery. Demand of labour is also not affected significantly by changes in the price of water or fertiliser.

TABLE 5.4 : Price Elasticities of Input Demand

Demand for Inputs	*Water*	*Price of Fertiliser*	*M. Capital*
Land	0.17	NS	NS
Water	–1.3	0.88	0.176
Fertiliser	0.40	0.374	0.078
Human Labour	NS	NS	NS
Modern Capital	0.196	0.19	NS

Note : NS stands for Not significant.

Before arriving at any policy conclusions from the above analysis one must remember, however, that this exercise pertains only to paddy cultivation. Any policy conclusions for agriculture as a whole would have to be based on a series of such studies for different crops. Sidhu and Baanante provide one such study for wheat in Punjab using the trans-log profit function approach.[27] They do not use the input classification that we do. A distinction is made between variable and fixed inputs and wheat supply functions depend on prices of variable inputs and amounts used of fixed inputs.

V. SUMMARY OF FINDINGS AND CONCLUSION

The methodology of our study enables us to comment on the nature of the production structure in paddy cultivation in Haryana in the late seventies and on the impact of aggregation of some inputs on the relationships of substitution and complementarity between them. The estimates of elasticities of substitution enable us to comment on a wide range of issues relating to paddy cultivation in the region.

The presence of a large number of significant parameters derived from a translog specification of the cost-function suggests that this is a better specification of the technology than the Cobb-Douglas.

The separability tests indicate that in studies on substitution possibility in paddy cultivation, it is crucial to treat water as a separate input for meaningful results to be obtained. With respect to modern capital, disaggregating it from conventional capital may provide greater insight. However, not doing so might not vitiate the estimates of other elasticities.

Paddy cultivation, we find, present considerable scope for substitution between inputs. The increase in the number and kind of inputs with the introduction of new varieties has been accompanied by an increase in the options open to the cultivator. In particular, there is scope for reproducible resources such as machinery and fertiliser to replace exhaustible resources like land and water. Such substitution will of course have a cost attached to it, particularly as machinery and fertiliser are both foreign exchange using inputs. Also, machinery (which in our case is mainly tractors) may also be substituting for human labour as well. It may, therefore be instructive to look at the own and cross price elasticities of input demand. These elasticities, with respect to the prices of purchased inputs turn up to be low (i.e. less than one) or insignificant with the one exception of the own-price elasticity of irrigation water.

The substitutability between water and fertiliser provides a pointer towards a two phase relationship between these two inputs, the first phase relevant for less water intensive crops marked by complementarity, the second for more water-intensive crops marked by substitutability.

The absence of significant relationships between labour and fertilizer and between water and labour seem to indicate that the employment effect of both these is not very substantial.

Input-use pattern by size-class seems to show more intensive use of fertiliser by large farms and labour by small farms. Water being divisible is unrelated to size. The indivisibility of modern capital is negated by the presence of a large market for hiring-of tractor services.

It must be remembered, in conclusion, however that the relationships between inputs and with reference to input demand that our study arrives at are crop specific and region specific. The stability of the relationships could perhaps be better established if the exercise were repeated for a few more years. This would probably be essential before far-reaching policy conclusions could be drawn.

Note : Reproduced from *Journal of Quantitative Economics*, Vol. I, No. 2, July, 1985; with kind permission of the publisher.

NOTES

1. *Cf.* Kopp and Smith (1982)
2. *Cf.* Dasgupta and Heal (1979).
3. Jaya and IRS yield on an average, 45 quintals of paddy p. hec. which sells at Rs. 80 p. qtl. (at 77-78 prices) to yield a gross output of Rs. 3600 p. hec. Basmati gives 30 qtls. per hect, which sells for Rs. 130 per qtl. to yield Rs. 3900 per hectare.
4. The coefficients of variation in input prices are as follows: land 24.80, human labour 51.70, conventional capital 45.72, modern capital 133.92, fertiliser 17.93.
5. For application to industry see Berndt and Wood (1975), Williams and Laumas (1981) and Uri (1980) among others. In agriculture, See Binswanger (1974), Kako (1978), Sidhu & Baanante (1980).
6. Sidhu and Baanante (1980).
7. *Cf.* Chopra (1982).
8. *Cf.* Fuss and Macfadden (1978) and Berndt and Wood (1975). For a good intuitive discussion see Humphrey and Moroney (1975).
9. It is there that the assumption regarding the behaviour of the cultivator, i.e. he minimised cost subject to given input prices, is required.
10. This last restriction, namely $\sum_{i=1}^{n} b_{iq} = 0$, has not been imposed in some studies (see Sidhu and Baanante (1980)).
11. See Allen (1938).
12. *Cf.* Kopp and Smith (1982).
13. See Berndt and Wood (1975).
14. This would mean that in that year farms selected for paddy in Punjab would be decided on keeping in mind the significance of paddy cultivation in them, *Cf.* Report of the Special Expert Committee on Cost of Production Estimates, Ministry of Agriculture, June 1980.
15. Moreover, fertiliser prices had registered a sudden increase at the beginning of that year and there was an unprecedented fall to fertiliser consumption, 1974-75 was also a drought year in many parts of the country.
16. This is similar to the method used in arriving at cost estimates for owned assets in many studies. See, e.g. NCAER, Implications of Tractorisation, (1980). The detailed data on asset structure available in the cost of cultivation studies enables us to use this method.
17. See p. 12 and 13 above.
18. 'Intensity of land Use' in the context of a single crop should be interpreted as the use of land in combination with higher levels of inputs to get higher yields. This would, however, lead to a higher cropping intensity as well, if the impact of the input-relation in the second crop is not in the reverse direction.

19. See Binswanger (1974) and Kako (1978).
20. See Note 18 p. 15.
21. *Cf.* Raj Krishna (1975) and Agarwal (1983).
22. This has been postulated for Ishikawa (1967). In discussing the relationship between irrigation and fertiliser in increasing agricultural output. "In a higher level of irrigation than this, substitution between both inputs becomes more prevailing" (footnote on p. 86 Ishikawa (1967). We maintain that in the case of paddy, a water intensive crop, this range is the relevant one.
23. See, for example, Agarwal's (1983) results on fertiliser use on tractor and non-tractor farms.
24. For a discussion on tests of separability see Berndt and Wood (1975). Also Fuss, MacFadden and Mundlak (1978).
25. The degrees of freedom are (2.335)
26. See Kopp and Smith (1982).
27. See Sidhu and Baanante (1981), that the complementarity relationship between water and fertiliser substantiates our 'two phase' hypothesis.

REFERENCES

Allen, R.G.D. (1938). *Mathematical Analysis for Economists.* London, Macmillan and Co.

Agarwal, B. (1983). *Mechanisation in Indian Agriculture: An Analytical Study based on the Punjab.* Allied Publishers Pvt. Ltd.

Berndt, E.R. and Wood, D.C. (1975). Technology prices and derived demand for energy. *Review of Economics and Statistics,* 57, 259-68.

Berndt, E. R. and Christenten, L. R. (1973). The internal structure of functional relations, separability, substitution and aggregation. *Review of Economic Studies,* 40, 403-410.

Binswanger, H. (1974). A cost function approach to the measurement of elasticity of factor demands and elasticities of substitution. *American Journal of Agricultural Economics,* 56, 377-386.

Binswanger, H. (1978). *The Economics of Tractors in South Asia : An Analytical Review.* ICRISAT.

Chopra, Kusum, (1974). Tractorisation and change in factor inputs, a case study of the Punjab. *Economic and Political Weekly* IX, (52).

Chopra, Kanchan (1982). Alternative sources of irrtigation and land use patterns. *Indian Journal of Agricultural Economics.* XXXVII (2), 171-183.

Dasgupta, P. and Heal, G. M. (1979). *Economic Theory and Exhaustible Resources.* Cambridge, Mass.

Chirstensen, L.R., Jorgenson, D.W. and Lau, L.J. (1973). Transcendental Logarithmic Production Functions. *Review of Economics and Statistics* 52, 287-300.

Fuss, M., Mac Fadden, D. and Mundlak, Y. (1978). A survey of functional forms in the economic analysis of production. In M. Fuss and MacFadden (eds.), *Production Economics. A Dual Approach to Theory and Application.* North-Holland.

Governmet of Punjab (India) (1969-72). Economics of Tractor Cultivation and Economcs of Production and Cultivation Practices of HYV's of Wheat, Maize and Paddy in Punjab, 1969-70 to 1971-72 Combined Report.

Hanumantha Rao, C.H. (1975). *Technological Change and Distribution of Gains in Indian Agriculture.* The Macmillan Company of India Ltd., Delhi.

Humphery, D.E. and Moroney, J.R. (1975). Substitution among capital, labour and natural resources products in American manufacturing. *Journal of Political Economy.* 83, 57-82.

Ishikawa, S. (1967). *Economic Development in Asian Perspective.* Kinokuneiya.

Kahlon, A.S. (1976). Impact of mechanisation on Punjab Agriculture with special reference to tractorisation. *Indian Journal of Agricultural Economics.* 31, 54-30.

Kako, T. Decomposition of derived demand for factor inputs, the case of rice production in Japan. *American Journal of Agricultural Economics.*

Kaneda, H. (1982). Specification of production functions for analysing technical change and factor inputs in agricultural development. *Journal of Development Economics* 11, 97-108.

Kopp, J. and Smith, V.K. (1982). The perceived role of materials in neoclassical models of production tecnology. In V. Kerry Smith and J.V. Kurtilla (eds), *Explorations in Natural Resource Economics.* John Hopkins.

Krishna, Raj (1975). Measurement of the direct and indirect employment effects of agricultural growth in technical change. In Edgar O. Edwards (ed.), *Employment in Developing Nations.* Columbia University Press.

Ministry of Agriculture (1980). *Report of the Special Expert Committee on Cost of Production Estimates.*

McIkerney, J. P. and Donaldson, G. F. (1975). The Consequences of Farm Tractors in Pakistan. *World Bank Staff Working Paper No. 210.*

Nadiri, M.J. (1982). Producers Theory. In K.J. Arrow and M.D. Intrilligator (eds.), *Handbook of Mathematical Economics.* North-Holland.

6

OUTPUT SUPPLY AND INPUT DEMAND EFFECTS OF HIGH YIELDING RICE AND WHEAT VARIETIES IN NORTH INDIAN AGRICULTURE

ROBERT E. EVENSON

The "Green Revolution" in South Asia, at least in its public visibility, is now history. A number of studies of the green revolution period have sought to measure the economic consequences for small and large farmers and for employment of labour and other inputs. It appears to be a general consensus of these studies that the green revolution was approximately "neutral" in terms of its relative effects on the employment of factors or production even though it was not neutral in its effect on output supply.

In this paper I examine the question of the neutrality of the green revolution effect on North Indian agriculture. The methodology employed is based on the proposition that Indian farmers are cost and profits conscious and that they alter production plans from season to season in a cost-minimizing fashion. Given this behaviour, decisions to supply different commodities and to employ different variable factors are interrelated because they are made jointly. Accordingly, a system of output supply and factor demand equations is utilised to estimate the consequences of price changes and of changes in the availability of high yielding varieties to farmers. These estimates indicate that the availability of the green revolution technology was decidedly non-neutral with respect to its effect on both the supply of outputs and the demand for inputs.

Methodology

In this paper I use an approach based on systems of output supply and factor demand equations derived from a maximized variable profits function.

Suppose these are n commodities, Y_i of which the first q are outputs and those indexed $q+1 \ldots n$ are variable inputs under the control of the individual agents, i.e., we have a vector of commodities Y such that

$$Y_i \geq 0 \text{ for } i = 1 \ldots m \text{ and } Y_i \leq 0 \text{ for i} = m + 1 \ldots n \quad (1)$$

These commodities have prices $P_i \geq 0$ for all i. π is variable profits or return to fixed factors of production and $\pi = Y'P$. Since inputs are negative quantities they subtract from revenues of the positive outputs. There are also k fixed factor of production, Z_k, $k = 1 .. k$ such as land, fixed capital, land quality and green revolution technology availability. If a sufficiently "well behaved" transformation function[1] exists $g\ (Y, Z) = 0$ and agents maximize variable profits π, then a profit function exists which relates maximized profits π^* to the prices of the variable commodities and the fixed factors.

$$\pi^* = \pi^*\ (P, Z) \quad (2)$$

It has the following properties (where π_i^* and π_{ij}^* are derivative in of the profit function with respect to the subscripts) :

(i) The profit function is monotonically increasing in P if it is an output price and monotonically decreasing in P if it is an input price.
The output supply and factor demand curves are

$$Y_i = \pi_p^*\ (P, Z) \begin{matrix} \geq 0 & i = 1, \ldots m \\ \leq 0 & i = m+1, \ldots n \end{matrix} \quad (3)$$

(ii) The profit function is *symmetric,* i.e.,

$$\pi_{ij} = \pi_{ji} \quad (4)$$

(iii) The profit function is convex, i.e., the (singular) matrix of its cross derivative π_{ij} is *positive semi-definite* or all its characteristics roots are positive or zero.

(iv) The profit function is homogeneous of degree one in prices and the supply and demand equations are homogeneous of degree zero in prices.

The matrix $[\eta_{ij}] = \left[-\frac{\partial Y_i}{\partial P_i} - \frac{P_j}{Y_j} \right]$ defines the factor demand and output supply elasticities and the following constraints hold :

$$\sum_{j=1}^{n} \eta_{ij} = 0 \tag{5}$$

In this study I specify (2) to have a normalized quadratic functional form[2]. This function is derived by stating the initial profit maximizing problem in terms of normalized prices $q_i = \frac{P_i}{P_n}$ where all prices and profit are divided by the price of the n'th commodity. Normalized profits then are written as

$$\bar{\pi} = \frac{\pi}{P_n} = \sum_{i=1}^{n-1} Y_{iqi} + Y_n \tag{6}$$

Shephard's Lemma then reads that $\frac{\partial \tilde{\pi}^*}{\partial q_i} = Y_i$. The quadratic normalized profit function is written as:

$$\bar{\pi}^* = a_o + \sum_{i=1}^{n-1} a_i q_i + \frac{1}{2} \sum_{j=1}^{n-1} b_{ij} q_i q_j + \sum_{i-1}^{n-1} \sum_{k} b_{ik} q_i Z_k \tag{7}$$

The output supply and factor demand curves for the first n – 1 factors are defined:

$$Y_j = \sum_{i-1}^{n-1} b_{ij} P_i / P_n + \sum_{k} b_{ik} Z_k \tag{8}$$

in terms of the original prices. Homogeneity of degree zero is imposed on the equations and cannot be tested. Symmetry is tested by imposing the restriction $b_{ij} = b_{ji}$, $i \# j$, $i \# n$.[3]

Data

In this study a set of District Level data for two North Indian regions is utilized. Data are in the form of annual measures of output and input per farm for the period 1959 to 1975. Table 6.1 provides a description of the data for each of the two regions. Four outputs, wheat rice, coarse cereals, and an aggregate of other crops, have been defined. Six "Z" variables which are presumed to be exogenous to the farmers decision process in the short run are defined. These include natural resource variables (rainfall), public goods variables, the availability of electricity, roads and public sector irrigation and the net cropped area land base. The green revolution impact is measured as the percentage of gross cropped areas of rice and wheat planted to high yielding varieties in the district.

TABLE 6.1 : Means of Variables

Variable	*North India Wheat Region Mean*	*Eastern Rice Region Mean*
Q Wheat	20678.190	10124.761
Q Rice	4319.350	22083.352
Q CRSCER	5660.205	4467.060
Q OCROPSB	25833.998	16114.353
Q FERT	4155.175	2641.756
QL (Labour)	22006.250	41818.244
QB (bullock)	21841.103	50139.948
QT (tractors)	1038.043	256.735
P Wheat	2.215	2.291
R Rice	2.057	1.879
P Crscer	2.174	2.390
PO Crops	2.898	3.288
P Fert	1.278	1.307
PL (wage)	2.041	2.111
PB	1.790	1.371
PT	1.577	1.577
Electricity	38.992	15.248
Roads	2.083	1.109
Rainfall	796.532	1133.998
HYVI	10.785	7.091
Irrigation Intensity	40.071	25.313
Net Cropped Area	1299.032	1711.519

Four variable inputs; fertilizers, bullock labour, tractor power, and labour are defined. It should be noted however, that only fertilizer quantity is measured on an annual basis at the district level. Tractor data are available for recent years in the North Indian Wheat region districts, labour and bullock labour variables are based on Census of Agriculture data and linear interpolation procedures.

Estimates

Tables 6.2 and 6.3 report estimates in elasticity form for the North Indian Wheat region based on district data from Punjab, Haryana, and Western Uttar Pradesh and the Eastern Rice region based on district data from central and eastern Uttar Pradesh and Bihar.

The estimation technique for these equations is as follows. In a first step each equation is estimated by Ordinary Least Square with year and region dummy variables to clear out year and region effects. In a second step, all equations are estimated jointly by Generalized Least Squares, taking account of error interdependence across equations (Zellner, 1963). Restrictions across equations are then tested via *F*-tests, and finally the system is reestimated with the full set of constraints. Elasticities are then computed using predicted values of those dependent variable which appear in the equations. The standard errors of the elasticities reported are approximations since they only take account of the variance of the coefficients in the formulae and not of the predicted values.

Each table reports an *F*-value for the symmetry restriction and the values of the characteristic roots of the Hessian matrix. The low *F*-values for the Eastern Rice results indicate that the imposition of the summetry restrictions does not produce a significant increase in the weighted mean square error for the system. The characteristic roots are all positive or, if negative, are insignificant. These statistics for the Northern Wheat data are not as satisfactory. They show that symmetry restrictions do result in a significant increase in mean square errors and that two of the characteristic roots are negative. Note however, that these estimates are based on variable which are in part interpolated.

The following generalizations can be made :

1. The basic output supply system is well estimated. All own price elasticities of supply are positive and significant in the North India System. The Eastern Rice system is less consistent with a *priori* expectations with insignificant wheat and coarse cereals elasticities. The rice supply equation, however is well estimated as are the other crops equations. The elasticity estimates are remarkably consistent for the two regions.
2. Both sets of data show wheat and rice to be good substitutes on

TABLE 6.2 : North India Wheat — Full System Elasticities

	Wheat	*Rice*	*Coarse Cereals*	*Other Crops*	*Fertilizers*	*Bullocks*	*Tractors*	*Labour*
Price of Wheat	.3180	–.3601	.3227	–0.652	.0469	.0376	–.0783	.0121
	(3.305)	(2.133)	(2.089)	(1.241)	(.2693)	(1.943)	(.3210)	(.3290)
Price of Rice	–.0706	.3031	–.0742	.0325	–.0797	.0287	.1212	.0501
	(2.133)	(2.335)	(.8711)	(1.641)	(.7821)	(2.271)	(.7714)	(1.997)
Price of Coarse Cereals	.0867	–.1016	.3507	–.0509	–.0035	.0147	.1266	–.1073
	(2.089)	(.8711)	(2.456)	(1.901)	(.0294)	(.9868)	(.6839)	(3.528)
Price of Other Crops	–.1093	.2780	–.3174	.1613	.4846	.0184	–.1135	.0237
	(1.241)	(1.641)	(1.901)	(2.141)	(2.884)	(1.007)	(.4790)	(.6995)
Price of Fertilizer	–.0055	.0478	.0016	–.0341	.1813	–.0540	–.6514	–.2008
	(.2693)	(.7821)	(.0294)	(2.884)	(1.617)	(5.233)	(2.490)	(4.264)
Price of Bullocks	–.0332	–.1289	–.0483	–.0097	–.4045	–.0177	–.7523	.1193
	(1.943)	(2.271)	(.9818)	(1.007)	(5.233)	(1.181)	(4.797)	(4.258)
Price of Tractors	.0027	–.0215	–.0164	.0024	–.1928	–.0297	1.4622	.0973
	(.3210)	(.7714)	(.6839)	(.4790)	(2.490)	(4.797)	(4.475)	(1.652)
Wages	–.1888	–.0167	–.2188	–.0364	.0323	.0020	–.1146	.0056
	(3.037)	(.1577)	(2.148)	(.6921)	(.2975)	(.1796)	(.7067)	(.2592)
Electricity	–.0311	.0039	.0039	.0592	.0362	.0122	–.1094	.0053
	(1.012)	(.0736)	(.0798)	(2.132)	(.6833)	(2.309)	(1.4776)	(.5632)
Roads	–.0138	–.0716	.2460	–.3304	–.5015	–.0902	1.6687	–.3321
	(.1270)	(.3951)	(1.423)	(3.374)	(2.725)	(4.630)	(6.373)	(9.566)

TABLE 6.2 (*Contd.*)

	Wheat	*Rice*	*Coarse Cereals*	*Other Crops*	*Fertilizers*	*Bullocks*	*Tractors*	*Labour*
Rain	.0360	.2126	–.3057	.0835	.5860	.0008	.4159	.0364
	(.4603)	(1.609)	(2.450)	(1.177)	(4.420)	(.0637)	(2.245)	(1.545)
High-Yielding Varieties	.2096	.4116	–.0339	–.0220	.5985	.0274	.3458	.0882
	(7.741)	(8.644)	(.8082)	(.8932)	(12.393)	(5.503)	(5.048)	(10.005)
Irrigation	1.4018	.8136	.8725	.7903	2.8267	–.0617	3.9959	.1623
	(10.454)	(3.611)	(4.276)	(6.481)	(11.893)	(2.474)	(11.687)	(3.627)
Net Cropped Area	–.1411	–.0501	.5418	.3078	–.7149	–.1940	–2.9649	.0192
	(.4999)	(1.862)	(1.329)	(1.190)	(1.484)	(3.634)	(4.170)	(2.038)

F-Value : 3.7380

Weighted R–Square : .5653

Characteristics Roots of the Hessain Matrix :

Col. 1	Col. 2	Col. 3	Col. 4	Col. 5	Col. 6	Col. 7
634.531	3467.02	1674.45	855.072	634.531	–2060.9	–3020.7

TABLE 6.3 : Eastern Rice — Full System Elasticities

	Wheat	*Rice*	*Coarse Cereals*	*Other Crops*	*Fertilizer*	*Bullocks*	*Tractors*	*Labour*
Price of Wheat	.0961	–.1377	.2661	.0899	–.2571	–.0099	–.0966	–.0170
	(.3517)	(1.443)	(1.044)	(.9941)	(.8967)	(.6753)	(.2486)	(.2670)
Price of Rice	–.2427	.3046	–.2130	–.0725	.0634	.0072	–.2004	–.0978
	(1.443)	(2.787)	(1.162)	(1.012)	(.3006)	(.6563)	(.7166)	(2.102)
Price of Coarse Cereals	.1229	–.0558	–.0081	–.0296	–.4100	.0065	.1495	.0924
	(1.044)	(1.162)	(.0371)	(.6283)	(2.082)	(.6262)	(.5147)	(1.975)
Price of Other Crops	.2091	–.0956	–.1489	.1923	1.2052	.0367	1.0045	.0292
	(.9941)	(1.012)	(.6283)	(1.670)	(4.338)	(2.487)	(2.590)	(.4658)
Price of Fertilizer	.0400	–.0056	.1380	–.0806	–.2832	–.0111	–.2699	.0208
	(.8967)	(.3006)	(2.082)	(4.338)	(1.553)	(1.582)	(.6540)	(.2614)
Price of Bullocks	.0295	–.0122	–.0422	–.0472	–.2138	–.0389	–.7033	.1042
	(.6753)	(.6563)	(.6262)	(2.487)	(1.582)	(4.851)	(3.174)	(2.961)
Price of Tractors	.0018	.0021	.0061	.0081	–.0327	–.0044	.1089	–.0679
	(.2486)	(.7166)	(.5147)	(2.590)	(.6540)	(3.174)	(.2465)	(.7899)
Wage	–.2567	.0001	.0142	–.0443	–.0718	0.139	.0073	–0.638
	(2.034)	(.0015)	(.1228)	(.7995)	(.5759)	(2.173)	(.0435)	(2.352)
Electricity	–.0544	.0038	.0081	.0038	.4254	.0027	–.0045	–.0393
	(1.371)	(.1500)	(2.343)	(.2301)	(10.909)	(1.393)	(.0830)	(4.781)
Roads	.3711	–.4038	–.0229	.0227	–.1228	–.0127	–1.3893	.1530

TABLE 6.3 (*Contd.*)

	Wheat	*Rice*	*Coarse Cereals*	*Other Crops*	*Fertilizer*	*Bullocks*	*Tractors*	*Labour*
	(1.166)	(1.841)	(.0968)	(.1515)	(.3819)	(.5370)	(2.485)	(1.618)
Rain	.5267	.2350	.0040	.1689	.5121	–.0078	.4994	.0851
	(2.997)	(2.091)	(0.263)	(22.382)	(3.071)	(.9351)	(2.270)	(2.453)
High Yielding Varieties	.6111	.1238	–.0926	–0.297	.1317	–.0006	.4342	.0567
	(9.331)	(3.064)	(1.878)	(1.201)	(2.322)	(.1805)	(5.604)	(4.281)
Irrigation	.7137	.0575	–.1254	–.2097	1.1886	0.790	1.3684	.2204
	(2.744)	(.3235)	(.6084)	(1.740)	(4.749)	(5.407)	(3.581)	(3.658)
Net Cropped Area	.1467	1.3594	1.2325	.5496	.4516	.1317	2.0343	–.2016
	(.2411)	(3.2350)	(2.646)	(1.926)	(0.770)	(3.780)	(2.275)	(1.420)

F-Value : .8692

Weighted R–square : .5269

Characteristic Roots of the Hessian Matrix :

Col. 1	Col. 2	Col. 3	Col. 4	Col. 5	Col. 6	Col. 7
9026.25	5007.02	2165.89	845.899	–70.94	–898.61	–1565.2

the supply side with wheat and coarse cereals showing a complementary relationship. It should be noted that crop rotations figure in this result. It is generally possible to phase coarse cereals in as a second crop with wheat. Only recently has this been possible to some extent with rice.

3. The fertilizer equation is only moderately well estimated. The own demand elasticities are negative as expected, but not all output price effects are positive.
4. The wage effects are generally as expected with higher wages causing a reduction in output supplied and a reduction in fertilizer demanded in both data sets.
5. The land variables, irrigation and net cropped area are generally consistent with expectation when taken together. The irrigation intensity measure is picking up a coefficient which reflects the added value of irrigated acreage. These data are indicating that irrigated acreage is very important in the North India data set particularly for wheat. It is possible that this variable is picking up some of the high yielding variety or green revolution effect.
6. The Z variables, electricity, roads and rainfall are generally consistent with expectation. Note that when one has several supply equations a public good type variable may have positive effects on one supply equation and negative on others. For example, in North India districts with high rainfall produce or supply more wheat and rice, but less of other crops. They also use more fertilizer.
7. The high yielding variety variable, which I am treating as a green revolution variable shifts the supply of wheat and rice of the right in both regions. This is as expected. It is also expected that the green revolution shifted the supply of coarse cereals and other crops (pulses, etc.) to the left. The data bears this out. We also expect it to shift the demand for fertilizer to the right.
8. The high yielding varieties and irrigation shift effects in the factor demand equations must be interpreted subject to the fact that we have not estimated consistent price effects in these equations. Experience with these systems indicates that Z variable effects are not highly sensitive to restrictions and price effects and probably not to errors in variables in price effects. I would accordingly wish to argue that the high yielding variety and irrigation effects are not likely to be seriously biased the these equations.

These estimates clearly do not show the green revolution to have been neutral with respect to its effect on factor demands. It has clearly had larger effects on fertilizer and tractor demand than on bullock labour demand and human labour demand. Indeed it may well have reduced the demand for bullock labour. This interpretation is strengthened by the consistency between the two data sets and effect of both the irrigation and high yielding varieties. (It appears that an interaction term for these two variables would have yielded stronger effects.)

Concluding Comments

A number of studies have attempted to obtain estimates of differential or biased effects on output supply and factor demand of green revolution and related technology. The methodology employed in this paper allows the analysis of such biases to be undertaken in consistent fashion. It also enables a number of other elasticity estimates of policy relevant parameters.

The actual estimates obtained are highly consistent with a *priori* expectations derived from the theory of profit maximization. This it the case even where data inadequacies are apparent. The estimates show that the green revolution in India has been decidedly non-neutral in its effect on both output supply and factor demand. It has favoured the supply of wheat and rice at the expense of other crops on the output side. It has also favoured the demand for fertilizer, tractors (and probably irrigation sewers) at the expense of bullock and human labour on the input side.

Note : Reproduced from *Indian Journal of Quantitative Economics,* Vol. I, No. 1, 1985, Punjab School of Economics, G.N.D. University, Amritsar with kind permission of the publisher.

NOTES

1. For the conditions which must be imposed on the transformation see Diewert. (1978).
2. For a discussion of normalized profit function see Lau, (1977).
3. Note that in this system we do not include the n'th commodity equation which has to be derived from equation (6) and the commodity equations.

REFERENCES

Bardhan, Kalpana, (1977), "Rural Employment Wages and Labour Markets in India : A Survey of Research", *Economic and Political Weekly,* Vol. 12, No. 26, 27, 28; June 26, July 2 and July 9, pp. 1012-1030, 1062-1074. 1101-1118.

Bardhan, Pranab K., (1976), *Determinants of Supply and Demand for Labour in a Poor Agrarian Economy : An Analysis of Household Survey Data in West Bengal,* Berkeley : University of California, Dept. of Economics Working Paper No. 83.

Bardhan, P.K., (1979), "Determinants of Supply and Demand for Labour in a Poor Agrarian Economy : An Analysis of Household Survey Data in Rural West Bangal." Paper presented at the *ADC-ICRISAT Conference on Adjustment Mechanism of Rural Labor Markets in Developing Areas,* ICRISAT, Hyderabad, India, August, 22-24.

Bardhan, P.K. and T.N. Srinivasan, (1974), *Poverty and Income Distribution in India,* Calcutta : Statistical Publishing Company.

Binswanger, Hans P., (1975), *The Use of Duality Between Production, Profit and Cost Functions in Applied Econometric Research,* Economics Department, International Crops Research Institute for the Semi-Arid Tropics, Occasional Paper No. 10.

Binswanger, Hans P., (1989), *The Economics of Tractors in South Asia : An Analytical Review,* New York : Agricultural Development Council.

Binswanger, H.P., (1978). "Risk Attitudes of Rural Household in Semi-Arid Tropical India", *Economic and Political Weekly,* Vol. XIII, No. 25, Review of Agriculture, June 24.

Diewert, W.E., (1973), "Functional Forms for Profit and Transformation Functions", *Journal of Economic Theory.* 6 : 284-316.

Evenson, R. E., (1972), Labour in the Indian Agriculture Sector, A Report to A.I.D. Economic Growth Center, Department of Economics, Yale University.

Krishna, Raj, (1975), *Unemployment in India,* Agricultural Development Council Reprint Series.

Lau, L.J. (1969), *Some Application of Profit Functions,* Memorandum No. 8-A and 86-B, Research Center in Economic Growth, 443, California : Encina Hall, Stanford University, Stanford.

Sidhu, S.S., (1974), "Relative Efficiency in Wheat Production in the Indian Punjab", *American Economic Review,* Vol. 64, September, pp. 742-751

Yotopoulos, P.A. and Larry J. Lau, (1973). "A Test of Relative Economic Efficiency", *American Economic Review,* Vol. 63, March pp. 214-223.

Zellner, A., (1963), "Estimates for Seemingly Unrelated Regression Equations : Some Exact Finite Sample Results", *Journal of the American Statistical Association,* Vol. 58, pp. 977-992.

7

FARM SIZE, IRRIGATION AND INTENSITY OF LAND USE IN INDIAN AGRICULTURE

G. K. CHADHA & R. K. SHARMA

The Issue

Rising land productivity is the only solid and enduring basis for Indian agriculture to achieve a continuous increase in production. A rise in land productivity can take place in many ways. One such possibility lies in increasing multiple cropped area, either by introducing new crops or by expanding area under existing ones. The underlying assumption is that every unit of the incremental cropped area gives some additional output, *albeit* small, so that output per acre of net sown area rises although it may or may not give an increase in output per acre of gross cropped area. It is, therefore, in the fitness of things that the future thrust in Indian agriculture goes, *inter alia,* in the direction of increasing multiple cropping. The National Commission of Agriculture (1976) has well recognised the importance of this fact and has given statewise projections of cropping intensity for 2000 and 2025 (Government of India, 1976, pp.150-153). In this connection, it is relevant to pose the question how are the prospective gains of rising cropping intensity going to be shared by various farm size groups? In particular, what is the future of small farms in the matter of cropping intensity? For locating a meaningful answer to this question, we must look to the experience of the past and the interim charges that are taking place at present.

The farm management studies (f.m.s.) for the fifties and the sixties, conducted for a few states, provided evidence to an inverse relationship between farm size and cropping intensity (Sau, 1978b, p. 104 and Bharadwaj, 1974, pp 94-95). This pointed to the fact that large farmers were not utilizing land, a very *scarce* and *non-reproducible national resource,* as optimally as small farmers. A higher cropping intensity on small farms provided a powerful explanation for the well known inverse relationship between farm size and productivity per acre, as also larger employment potential on such farms (Bharadwaj, 1974). On this basis, a strong case for land redistribution was also made. If land is transferred, through a related tenancy arrangement or otherwise, from those who keep it idle for a good part of the year to those who keep on cultivating the same for a much larger part of the year, it will obviously generate higher output and greater employment per unit of land area. Ranjit Sau says forthrightly that "In India, whoever has the ability but not the willingness to utilize the land properly, should have no right to cultivate the land, despite his ownership. Whoever has the willingness, but not ability (as in the case of peasants and landless labourers) should receive land as well as other support from the state. Such a policy may impart dynamism to the rural sector" (Sau, 1978a, p. 1233).

A tendency has somehow persisted among the researchers to accept the inverse farm size-cropping intensity relationship almost as a universal phenomenon for Indian agriculture. We have reasons to believe that this is not so. Firstly, the inverse relationship for the fifties was statistically non-significant in many cases (Bharadwaj, 1974, p. 95). Secondly, since the number of sample districts for each selected state generally did not exceed two, for the fifties as well as the sixties, generalizations for the state as a whole were not in order. Thirdly, some of the districts being common between the f.m.s. of the fifties and the sixties, the net evidence relates to a very limited spatial coverage across just a few states. This is obviously a big research gap. It is a pity that there exists not a single study, for any of the three decades, which encompasses all districts of even a single state, let alone all districts of all states. We wish to point out that the gap for the seventies is especially distressing because the recent technological changes have affected, *inter alia,* cropping intensity differently in different states and possibly, different farm size groups in each state. One may like to see if the inverse relationship between farm size and cropping intensity has ceased to operate in some areas although some fellow researchers strongly hold that it still persists inside as well as outside HYV areas, and, on irrigated as well as unirrigated lands (Sau, 1978a. p. 1281).

This paper seeks to examine whether, after the arrival of the Green Revolution, there exists an inverse farm size cropping intensity relationship for Indian agriculture as a whole. The canvas of our analysis is very wide. We examine the relationship separately for each of the 318 districts of India, spread over almost all states.[1] The merit of working out districtwise relationship needs no emphasis. Within a district, spatial variation in soil types and land quality, irrigation structure, cropping patterns, etc. are fairly limited, compared with the state as a whole. In fact, one would like to use data even at the tehsil or lower levels but this is only a wishful thinking if one is covering the whole of India.

The Data and the Methodology

The 1970-71 Agricultural Census Reports for various states have been used extensively for the present paper. For the first time, district-wise tables on various aspects of agrarian economy have become available for various farm size groups.[2]

We have run weighted regressions between (mean) cropping intensity (Y) and (Mean) farm size (X), separately for all the 318 districts. In the first instance, the data scatters for all the districts were scrutinised to decide the type(s) of statistical fits. In most cases, the scatters suggested linerar fits. Only in respect of Kerala districts as a whole and a few others, the scatters suggested quadratic fits.

The net operational area reported by a farm size group was divided by the number of holdings in that group to arrive, as its mean farm size (X). Similarly, gross cropped area reported by a farm size group was divided by its net sown area to derive the cropping intensity for that group.[3] The number of holdings in each farm size group was used as its weight.[4]

A set of three regressions was computed for those districts which had 10.0 per cent or more of their net sown area as irrigated (Irrigation Base) : set I for overall cropping intensity (Y_1 = Gross Cropped Area divided by Net Sown Area); set II for cropping intensity on irrigated areas (Y_2= Irrigated Gross Cropped area divided by Irrigated Net Sown Area); and set III for cropping intensity on unirrigated areas (Y_3 - Unirrigated Gross Cropped Area divided by Unirrigated Net Sown Area). For districts with an Irrigation Base less than 10.0, only the overall cropping intensity was estimated.

The formal expressions for the linear regression equations are :

$$Y_i = a_i + b_i \; X \qquad ..(1)$$

i = 1 for overall cropping intensity;

i = 2 for cropping intensity on irrigated areas;

i = 3 for cropping intensity on unirrigated areas.

Likewise, the formal expressions for the quadratic equations are :

$$Y_i = a_i + b_i \; X + c_i + X^2 \qquad ...(2)$$

i = 1, 2, 3 has the same meaning as in equation (1) above.

In all, we have 318 weighted regressions between Y_1 and X, 197 between Y_2 and X, and 195 between Y_3 and X. For paucity of space, only a summary of the results is given in Table 7.1.[5] Besides regressions, weighted correlation coefficients were also worked out between farm size and either of (i) Irrigation Base (percentage of Net Sown Area irrigated) and (ii) Improved Irrigation Base (percentage of net irrigated area contributed by tubewells and wells). These coefficients are designated as r_{12} and r_{13} respectively. While r_{12} was computed only for those districts (198 in all) which had their Irrigation Base higher than 10.0, r_{13} was estimated for those greater than ten per cent irrigation base districts (146 in all) which had their improved irrigation base greater than 10.0 per cent.

Before discussing the results of Table 7.1, we better sort out a few points. Firstly, it was considered advisable to exclude the last farm size group (i.e. 50.0 hectares and above), for two good reasons. The mean farm size of this open-end group was miles away from the mean farm size of the preceding group (i.e. 40.0 – 50.0 hectares), thus creating a terribly big gap between the last and the penultimate values of the variable X. Also, the number of holdings in this group was practically nil (in many districts, no holdings at all) compared with many times more in smaller groups. Inclusion of such distantly located few farms should have served no purpose. On the contrary, we might have run the risk of introducing spurious kink or flatness in the farm size cropping intensity relationship that otherwise seemed to run smoothly between the smallest farm size group and the 40.0 – 50.0 hectare group.

Secondly, the quadratic fits were estimated strictly when the data scatters so dictated.[6] In those cases, we estimated linear fits also but do not include them in Table 7.1, to avoid jumble.

Thirdly, some aggregation bias in the estimates of b's cannot be ruled out precisely because our regressions are based on group means.

We, however, believe that the weighted regressions would keep such a bias to its minimum. Also, in respect of some districts, the cropping intensity was noticed either as constant or niggardly increasing up to, say, the second farm size group, but the estimated relationship was imposingly inverse. In such case, the estimational bias should affect slightly, if at all, the value of the intercept and should have no effect whatsoever on the direction of the regression coefficient. To put the record straight, we are immediately interested in the direction of the regression coefficients rather than their sheer magnitudes.

Lastly, the estimated b's are of three types (i) negative and statistically significant; (ii) positive and statistically significant; and (iii) statistically non-significant (whether positive or negative). We call types (i), (ii) and (iii) as definitely negative, definitely positive and neutral, respectively. Definitely positive b's suggest the relative edge of large farm size groups, definitely negative b's suggest the contrary situation and neutral b's imply that the cropping intensity is constant over the full range of farm sizes. Deviating from convention, we consider the significance of estimated b's up to 0.20 level, only to make sure that as many of them are assigned to definitely positive or definitely negative types as possible, and the neutral coefficients are indeed definitely neutral.

The Results

Table 7.1 gives a mixed lot of results. All districts of Kerala conform to a quadratic relationship, something resembling a flat, inverted U. The intensity keeps on rising till a certain farm size is reached beyond which it reverses its direction. It is an interesting situation for which we have two possible explanations. The first emerges from the perusal of cropping patterns pursued by different farm size groups. The percentage of cropped area under such crops as have a definite scope for promoting multiple cropping goes on increasing typically till about the middle farm size groups and reverses itself for higher groups. Rice is a typical example. On the other hand, the long-duration (but commercial) crops such as tapioca, coconut, cashew, mango, jack, etc. occupy a high priority in the cropping pattern of very small, small and medium farms, in that order. The second explanation which reinforces the first, emerges from the fact that in most districts of Kerala, there is extreme preponderance of very small and small farms. For example,

TABLE 7.1 : Results of the Regression Exercises Between farm size and cropping intensity

State	Regression co-efficients for											
	Overall (b_1)				*Irrigated areas (b_2)*				*Unirrigated areas (b_3)*			
	Total	*–ve*	*+ve*	*Neutral*	*Total*	*–ve*	*+ve*	*Neutral*	*Total*	*–ve*	*+ve*	*Neutral*
1. Andhra Pradesh	21	18	1	2	19	8	7	4	19	10	5	4
2. Assam	9	5	—	4	4	3	1	—	4	2	—	2
3. Bihar	31	18	6	7	19	7	3	9	19	11	3	5
4. Gujarat	19	14	1	4	9	4	14	4	9	8	1	—
5. Haryana	10	10	—	—	10	10	—	—	10	10	—	—
6. Himachal Pradesh	11	11	—	—	6	6	—	—	5	5	—	—
7. Karnataka	16	13	—	3	8	5	1	2	8	4	2	2
8. Jammu & Kashmir	9	7	—	2	8	6	1	1	7	6	—	1
9. Kerala	11	—	11	—	4	—	3	1	4	—	4	—
10. Madhya Pradesh	43	41	—	2	10	5	3	2	10	9	—	1
11. Maharashtra	26	18	3	5	9	5	1	3	9	9	—	—
12. Punjab	12	12	—	—	12	10	2	—	12	10	2	—
13. Rajasthan	26	23	—	3	15	13	1	1	15	14	—	1
14. Tamil Nadu	11	11	—	—	11	7	2	2	11	8	2	1
15. Uttar Pradesh	48	44	1	3	48	27	6	15	48	41	1	6
16. West Bengal	15	14	—	1	5	4	1	—	4	4	—	1
Total (excluding Kerala)	307	259	12	36	193	120	30	43	191	151	16	24
Percentage	(100)	(84.36)	(3.91)	(11.73)	(100)	(62.18)	(15.50)	(22.28)	(100)	(79.06)	(8.38)	(12.56)
Total (including Kerala)	318	259	23	36	197	120	33	44	195	151	20	24
Percentage	(100)	(81.45)	(7.23)	(11.32)	(100)	(60.90)	(16.75)	(22.34)	(100)	(77.44)	(10.26)	(12.30)

about 70.0 per cent of farms in Kerala are below 0.5 hectare each; in some districts, this percentage is as high as 85.0. On all farms, as also on these tiny farms, coconut cultivation is economically a must; it is almost an institution. As a matter of fact, 55 to 60 per cent of gross cropped area on farms below 0.5 hectares is occupied by two crops alone (coconut and tapioca), (Government of Kerala, 1973 p. 242). Cereals and pulses enjoy relatively low priority on such farms. To conclude, anything like inverse farm-size cropping intensity relationship is ruled out for Kerala. Let us, therefore, keep Kerala aside and consider the remaining 307 districts.

There is no doubt that definitely negative values of b_1 (Overall cropping intensity) and b_3 (cropping intensity on unirrigated areas) are given by an overwhelming majority and such values of b_2 (cropping intensity on irrigated areas) are given by a majority of districts, in each state. It is particularly interesting to see an almost conclusive evidence in favour of the inverse relationship in the Green Revolution areas such as the Punjab and Haryana. In brief, the 1970-71 evidence tilts heavily in favour of an inverse relationship. But this cannot hide the fact that a sizeable number of districts show neutral values of b_1 (36 out of 307 : 11.73 per cent), b_2 (43 out of 193 : 22.28 per cent) and b_3 (24 out of 191 : 12.56 per cent). Since none of these districts ever figured in the earlier f.m.s. studies, it is difficult to verify whether in their case a neutral relationship has persisted over time, or the inverse relationship has yielded to a neutral relationship. If we are to believe that these districts *indeed* had an inverse relationship in the past, it is clear then,in an *expost* sense, that the small farmers in these districts have lost their traditional edge of higher cropping intensity.

It may also be noted that some districts show definitely positive value of b_1, b_2 and b_3. Once again, considering all the 307 districts together, definitely positive values of b_1 are obtained in 12 districts (out of 307 : 3.91 per cent; Table 7.2), of b_2 in 30 (out of 193 : 15.54 per cent; Table 7.3) and of b_3 in 15 districts (out of 191 : 7.85 per cent; Table 7.4). Clearly, the percentage of such districts is fairly substantial for cropping intensity on irrigated areas (15.54) while it is very low for overall cropping intensity (3.91). More surprising is, however, the fact that about 8.0 per cent of the districts show a definitely positive relationship *even* on unirrigated areas. Although it is difficult to think of a general explanation, or to spell out the reasons in full, for a definitely positive farm size cropping intensity relationship for different areas, yet the disparate results can be reconciled to some extent. Between the irrigated and the unirrigated areas, the positive relationship on the former is more understandable, especially because of recent changes in agricultural technology including irrigation expansion/improvement. Let us look at the 30 cases of definitely positive b_2's (Table 7.3). Only in three cases (Visakhapatnam—Andhra Pradesh, Bojpur—Bihar and Varanasi—U.P.), a definitely positive b_2 has a definitely positive b_1 also. Out of the remaining 27 cases of definitely positive b_2's, 22 have definitely negative and five neutral b_1's. This means a definitely positive relationship on irrigated areas has at best limited effect, and is not automatically transmitted on to the overall relationship. It seems, large farms in those 30 districts are making a much more intensive use of their irrigated areas than the unirrigated areas. In terms of traditional arguments, this is not expected to happen because the f.m.s. studies showed an inverse relationship between farm size and irrigation base. Our data also show a negative relationship between farm size and irrigation base (r_{12}) in most of these cases, irrespective of the irrigation level of each district. What then is the real explanation? One possible explanation emerges of a distinction is made between irrigation level and irrigated structure. We argue that just as the level of irrigation was an important determinant of cropping intensity in traditional agriculture, the type of irrigation becomes so under the new technological conditions. Our contention is duly verified by the fact that whereas the correlation is duly verified by the fact that whereas the correlation between farm size and irrigation base (r_{12}) is negative in most of these 30 cases, the correlation between farm size and improved irrigation base (r_{13}) is positive in almost all these cases. It easily follows that assured availability of irrigation rather than the level of irrigation as such, has promoted higher cropping intensity on large farms in many cases. The argument cannot, however, be reversed since we find, for many other districts, a positive correlation between farm size and improved irrigation base (r_{13}) but a negative relationship between farm size and cropping intensity on irrigated areas. In these cases, either the size-biased effect of improved irrigation base are reduced through the widespread custom of hire/sale of tubewell water, or a more effective use of well irrigation, or because of reasons not related to irrigation.

TABLE 7.2 : 12 Cases of Definitely Positive b_1's

State	*District*	b_1	b_2	b_3	*Irrigation Base*	r_{12}
Andhra Pradesh	1. Vishakhapatnam	+ve	+ve	+ve	25.80	-0.21
Bihar	1. Bhojpur	+ve	+ve	N	64.44	0.57
	2. Dhanbad	+ve	NE	NE	3.98	NE
	3. Madhubani	+ve	N	+ve	11.26	0.33
	4. Ranchi	+ve	NE	NE	3.28	NE
	5. Saharsa	+ve	N	+ve	18.52	0.28
	6. West Champaran	+ve	N	+ve	29.69	0.44
Gujarat	1. Rajkot	+ve	–ve	+ve	14.09	–0.65
Maharashtra	1. Akola	+ve	NE	NE	0.57	NE
	2. Buldana	+ve	NE	NE	1.51	NE
	3. Kolaba	+ve	NE	NE	2.15	NE
Uttar Pradesh	1. Varanasi	+ve	+ve	+ve	59.94	–0.18

N = Neutral; NE = Not estimated

TABLE 7.3 : 30 Cases of Definitely Positive b_2's

State	*District*	b_1	b_2	b_3	*Irrigation Base*	r_{12}	r_{13}
Andhra	1. Anantpur	-ve	+ve	N	11.79	0.38	0.55
	2. Cuddapah	-ve	+ve	-ve	26.35	-0.56	0.50
	3. Chittoor	-ve	+ve	-ve	33.47	-0.52	0.33
	4. Ongole	N	+ve	-ve	19.12	-0.47	0.71
	5. Nellore	-ve	+ve	-ve	62.43	-0.49	0.57
	6. Visakhapatnam	+ve	+ve	+ve	25.80	-0.21	NE
	7. Warangal	-ve	+ve	N	24.77	-0.51	0.56
Assam	1. Kamrup	-ve	+ve	-ve	11.99	0.65	NE
Bihar	1. Bhojpur	+ve	+ve	N	64.66	0.57	0.29
	2. Gopalganj	N	+ve	-ve	33.41	-0.48	0.11
	3. Saran	-ve	+ve	-ve	23.61	-0.41	0.36
Gujarat	1. Gandhi Nagar	N	+ve	-ve	16.96	0.76	0.21
Jammu & Kashmir	1. Doda	N	+ve	-ve	10.98	-0.53	NE
Karnataka	1. Mysore	-ve	+ve	-ve	14.44	-0.35	0.69
Madhya Pradesh	1. Bilaspur	-ve	+ve	-ve	12.83	-0.59	NE
	2. Morena	-ve	+ve	-ve	24.65	0.73	NE
	3. Raipur	-ve	+ve	-ve	23.19	-0.57	NE
Maharashtra	1. Satara	-ve	+ve	-ve	13.47	-0.76	0.82
Punjab	1. Hoshiarpur	-ve	+ve	-ve	38.88	0.53	0.52
	2. Rupar	-ve	+ve	-ve	43.15	0.54	0.59
Rajasthan	1. Bhilwara	-ve	+ve	-ve	42.95	-0.06	0.01
Tamil Nadu	1. Chingleput	-ve	+ve	-ve	72.98	-0.65	0.55
	2. North Arcot	-ve	+ve	-ve	45.47	-0.56	0.55
Uttar Pradesh	1. Basti	N	+ve	-ve	58.45	-0.62	0.46
	2. Deoria	-ve	+ve	-ve	50.94	-0.63	-0.57
	3. Etah	-ve	+ve	-ve	62.76	-0.06	0.17
	4. Rai Bareily	-ve	+ve	-ve	44.45	-0.68	0.66
	5. Rampur	-ve	+ve	-ve	26.75	0.68	0.54
	5. Varanasi	+ve	+ve	+ve	59.94	-0.18	0.59
West Bengal	1. Murshidabad	-ve	+ve	-ve	22.71	-0.26	NE

N = Neutral; NE = Not Estimated

TABLE 7.4 : 15 Cases of Definitely Positive b_3's

State	District	b_1	b_2	b_3	*Irrigation Base*	r_{12}
Andhra Pradesh	1. Khammam	N	N	+ve	18.49	-0.44
	2. Medak	-ve	-ve	+ve	17.57	-0.46
	3. Nizamabad	-ve	-ve	+ve	43.39	-0.59
	4. Visakhapatnam	+ve	+ve	+ve	25.80	-0.21
Bihar	1. East Champaran	N	-ve	+ve	10.60	0.41
	2. Madhubani	+ve	N	+ve	11.26	0.33
	3. Saharsa	+ve	N	+ve	18.52	0.28
Gujarat	1. Rajkot	+ve	N	+ve	14.09	-0.65
Karnataka	1. Bangalore	N	-ve	+ve	12.61	-0.32
	2. Shimoga	-ve	-ve	+ve	45.64	-0.62
Punjab	1. Kapurthala	-ve	-ve	+ve	83.16	-0.61
	2. Ludhiana	-ve	-ve	+ve	90.97	0.16
Tamil Nadu	1. Coimbatore	-ve	-ve	+ve	35.13	-0.45
	2. Dharampuri	-ve	-ve	+ve	14.52	-0.48
Uttar Pradesh	1. Varanasi	+ve	+ve	+ve	59.94	-0.18

N = Neutral; NE = Not estimated

We must sum up the conclusions emerging in the preceding paragraphs. First, about the overall cropping intensity. There is undoubtedly a strong evidence in favour of an inverse farm size-cropping intensity relationship except in Kerala. Two points need to be stressed in particular. Firstly, the Green Revolution areas (Punjab, Haryana, Western U.P., etc.) conform to the inverse relationship almost completely and conclusively. This clearly suggests that the arrival of the new farm technology has not thrown the small farmers out of their traditional position of higher cropping intensity. Secondly, among the non-Green Revolution States, the cases of positive relationship are practically nil. Only six (out of 31) in Bihar and three (out of 26) in Maharastra are worth mentioning. However, there is a sizeable number of districts in the non-Green Revolution states where the relationship is neutral. If indeed these districts had an inverse relationship in the past, clearly then the small farmers there have lost their traditional edge.

We have also examined the cropping intensity on irrigated as well as on unirrigated areas. On irrigated areas, again a majority of districts suggest an inverse relationship but the percentage of districts showing a positive as also the neutral relationship is fairly high. In some districts, an improved irrigation structure seems to have promoted higher cropping intensity on irrigated areas of large farms, while for many others, the inverse relationship persists in spite of a positive correlation between farm size and improved irrigation base. On unirrigated areas, again an imposing evidence exists for the inverse relationship. To a very large extent, Sau's assertion that "the intensity of cropping varies inversely with farm size, in HYV areas and outside HYV areas, in irrigated lands and in unirrigated land, during the fifties as well as in the sixties and the seventies" stands proved (Sau, 1978a, p. 1281). Nevertheless, it is equally important to emphasize that the inverse relationship has been reversed in many Indian districts.

Given the crucial importance of irrigation in raising cropping intensity, it is essential to pursue the behaviour of r_{12} and r_{13} in a bit more detail. In most of the earlier f.m.s. studies, an inverse relationship (i.e. negative r_{12}) was observed between farm size and the proportion of net sown area irrigated. Thanks to the preponderant position of canal irrigation and its delivery system, and a near absence of tubewell irrigation, the small farmers could irrigate a relatively larger proportion of their net sown area. A relatively higher level of irrigation on small farms offered a formidable explanation for the inverse farm size-cropping intensity relationship. It is interesting to note that about 70 per cent of districts still show negative r_{12} (Table 7.5). In other words, for a big majority of districts, a negative relationship between farm size and irrigation base still persists. However, the reverse is true about the improved irrigation base in as much as about 77.0 per cent of districts show positive r_{13} (Table 7.5).[7] The investment in tubewell irrigation, on a wide scale, is a recent phenomenon and being size biased, the relative edge of large farms is quite understandable.

TABLE 7.5 : Matrix of Correlation Coefficients

State	r_{12}			r_{13}		
	Total	*+ve*	*–ve*	*Total*	*+ve*	*–ve*
1. Andhra Pradesh	19	—	19	12	12	—
2. Assam	4	3	1	—	—	—
3. Bihar	19	10	9	16	13	3
4. Gujarat	9	2	7	9	7	2
5. Haryana	10	5	5	8	6	2
6. Himachal Pradesh	5	—	5	—	—	—
7. Jammu & Kashmir	7	1	6	—	—	—
8. Karnataka	8	1	7	6	4	2
9. Kerala	5	5	—	—	—	—
10. Madhya Pradesh	10	1	9	6	2	4
11. Maharashtra	9	1	8	7	7	—
12. Punjab	12	6	6	12	10	2
13. Rajasthan	16	4	12	15	11	4
14. Tamil Nadu	11	—	11	11	10	1
15. Uttar Pradesh	48	18	30	44	30	14
16. West Bengal	6	4	2	—	—	—
Total	198	61	137	146	112	34
Percentage	(100)	(30.81)	(69.19)	(100)	(76.71)	(23.29)

Let us now pose the central issue. The issue is that in spite of a positive relationship between farm size and improved irrigation base, there exists a negative relationship between farm size and corpping intensity in a very big majority of districts. It means that in spite of a technological disadvantage in the matter of improved irrigation facilities, the marginal and small farmers continue to cultivate their land more intensively over a large part of Indian agriculture. Our hunch is that an asssured and flexible year-round supply of their family labour continues to play a critical role in increasing cropping intensity. Perhaps, their family labour will continue to play this role in future as well.

Note : Reproduced from *Artha Vijnana,* V. 24, No. 1, March 1982, with kind permission of the publisher.

NOTES

1. Because of non-availability of data, all Union Territories and the far eastern hill states could not be included in our study. Also because of some internal inconsistency of data (e.g. gross cropped area falling short of net sown area), we were obliged to leave out the State of Orissa and a few other districts. Our coverage gap is, however, very small in as much as about 95.0 per cent of net sown area (1970-71) of the country is covered by us.
2. Generally, 12 farm size groups have been provided for in most of these reports. These are (in hectares); upto 0.5; 0.5-1.0; 1.0-2.0; 2.0-3.0; 3.0-4.0; 4.0-5.0; 5.0-10.0; 10.0-20.0; 20.0-30.0; 30.0-40.0; 40.0-50.0; and above 50.0. For some states such as Andhra Pradesh, Jammu & Kashmir and Kerala, holding up to 1.0 hectare were divided into 3 or 4 group instead of only two.
3. Recently, an alternative measure of cropping intensity has been suggested. See, B. Dayal, 1978 "A Measure of Cropping Intensity", *The Professional Geographer,* Vol. XXX, August 1978, No. 8 pp. 280-96. This measure takes into account of not only the area under each crop but also the duration for which the crops stand in the field. In our case, the duration factor would really distort the farm size-cropping intensity relationship if the cropping pattern goes on changing at every step as we move up (or down) the ladder of farm size groups. This is most certainly the case in districts where plantation crops are important. We have taken care of this point by excluding such plantation crops as have definite bias for or against particular farm size groups as regards its effect on cropping intensity. Rubber, Coffee, Tea and their accompaniments (e.g. Cardamom, Pepper, etc.) grown on inter-culture basis.
4. In terms of regression equation $Y_i = a + b X_i$, the expression for weighted regression coefficient is—

$$\frac{\Sigma W_i (X_i - \overline{X}) (Y_i - \overline{Y})}{\Sigma W_i (X_i - \overline{X})^2}$$

against $\frac{\Sigma (X_i - \overline{X}) (Y_i - \overline{Y})}{\Sigma (X_i - \overline{X})^2}$ used in unweighted regression.

In the present study, the unweighted regressions are obviously meaningless.

5. Interested researchers may request for a copy of the 18-page supplement (mimeo) on districtwise regression and correlation results from the first author.
6. The scatters were unequivocally suggestive of the type of relationship involved between cropping intensity and farm size so that we do not feel the necessity of employing *F*-test to confirm (or reject) linearity. Interested researchers may consult Kanpur, J.C. & Saxena, H.C., 1967 *Mathematical Statistics*, S. Chand and Company p. 347 for such tests.
7. It is perhaps puzzling that for about one-fourth of the districts, the improved irrigation base should vary inversely with farm size. The puzzle is, however, because of our rather broad definition of 'improved sources of irrigation'. In our definition, both wells and tubewells are treated as improved sources of irrigation. In most of these puzzling districts, well irrigation constitutes a much larger part of the improved irrigation base, and does not seem to cause a size bias. For example, Gwalior, Shivpuri and Tikamgarh districts of Madhya Pradesh, show improved irrigation entirely through wells; in Partapgarh, Fatehpur, Jaunpur and Sultanpur districts of Uttar Pradesh, Siwan and Palemau of Bihar, and Morena of Madhya Pradesh, the ratio of well to tubewell irrigated area is roughly 15:1, 5:1, 3:1, 2:1, 6:1, 5:1 and 2.5:1 respectively and in Allahabad, Badaun and Bareily districts of Uttar Paradesh the ratio is roughly 1:1 and so on. It may not, therefore, be a surprise if the negative correlation between farm size and improved irrigation base, for most of these so-called puzzling districts actually irrigation base, for most of these so-called puzzling district actually become possitive if only we exclude well irrigation from our definition of improved sources. In that case, we are in a more comfortable position to conclude that for most districts in India, improved irrigation base is positively related with farm size.

REFERENCES

Bharadwaj, Krishna (1974), *Production Conditions in Indian Agriculture*, Cambridge University Press.

Dayal, B., (1978), "A Measure of Cropping Intensity", *The Professional Geographer*, Vol. XXX, No. 8, Aug.

Govt. of India (1976), *Report of the National Commission of Agriculture*, Vol. 5, Ministry of Agriculture and Irrigation.

Govt. of Kerala (1973), *World Census of Agriculture, 1970-71*, Vol. II, Report for Kerala State.

Kapur, J. N. and Saxena, H.C. (1967), *Mathematical Statistics*, S. Chand & Co.

Sau, Ranjit (1978a), "Growth, Employment and Removal of Poverty", *Economic and Political Weekly*, Special Number, Aug.

——(1978b), *Unequal Exchange, Imperialism and Underdevelopment*, Oxford University Press.

8

AN ANALYSIS OF STRUCTURAL CHANGES IN AGRICULTURE

A Case of Punjab

K. S. DHINDSA

A few attempts have been made at structural analysis of the primary sector. Where the primary activities are predominantly agricultural, varied interpretations have been ascribed to the term changing structure of agriculture. Prof. Bhalla (1) interprets it as changes brought about by technology in the distribution of income, consumption and investment, while Prof. Vyas (2) defines it as changes in the distribution of land (man-land relations) and changes in the size pattern of land holdings. However, a comprehensive definition of structural change in agriculture should include changes in the resource, structure of agricultural changes in the productivity of different crops, i.e., distribution of land holdings, changes in the composition of workers engaged in agriculture and percentage changes in contribution to the state's Net Domestic Product of different components of the primary sector, namely, agriculture, animal husbandry, forestry and fishery.

The two objectives of this paper are :[1]

- To study the changes in the resource structure of Punjab's agriculture during 1959-61 and 1969-71.
- To examine the changes in the production structure of Punjab's agriculture during 1959-61 and 1969-71.

Accordingly this paper has been grouped under four different sections:

Section I deals with the identification of the variables representing the structure of agriculture resources and that of production, while Section II defines the methodology. Section III presents the discussion of the results and the conclusions are given in the last section.

I. IDENTIFICATION OF THE VARIABLES REPRESENTING THE AGRICULTURAL RESOURCES STRUCTURE AND PRODUCTION STRUCTURE

Broadly speaking agricultural resource structure is represented by four resource inputs—labour, land, water and capital. The labour resource input is proximated by five variables:

(i) Workforce engaged in agriculture as a proportion of total male workers;
(ii) Male landless agricultural labourers per 100 cropped hectares;
(iii) Male cultivators per 100 cropped hectares;
(iv) Male labour force per 100 cropped hectares; and
(v) Percentage of literate workers amongst rural male workers.

The land resource input is represented by five variables :

(i) Gross cropped area;
(ii) Gross cropped area under the set of eleven selected crops;
(ii:) Size of cultivating units;
(iv) Proportion of holding units below five hectares to total holding units (%); and
(v) Proportion of holding units below 5 hectares to total holding units (%).

The water resource input is represented by two variables:

(i) Irrigated area in proportion to net sown area.
(ii) Number of pump-sets per 100 gross irrigated area.

The capital resource input is proximated by seven variables:

(i) Number of pair units of draft animals (working male cattle) per 100 cropped hectares.

(ii) Number of ploughs per 10 hectare unit of cropped area.
(iii) Number of sugarcane crushers[2] per 100 hectare unit of cropped area.
(iv) Number of carts per 100 hectare of cropped area.
(v) Number of tractors per lakh cropped area.
(vi) Fertilizer consumption (N+P+K) per cropped area.
(vii) Area under high-yielding varieties of foodgrain crops to total area under foodgrain crops.

As regards the estimation of agricultural productivity, we have taken sum total of the gross output value of eleven important crops—wheat, rice, maize, gram, bajra, sugarcane, potatoes, groundnut, rape, mustard, and cotton. To obtain the gross output value of individual crops triennial average prices (1969-71) are multiplied with the corresponding quantity figures. Estimates of productivity per worker are obtained by diving the sum total value of the set of eleven selected crops by the total male workers. Similarly estimates of productivity per hectare are obtained by dividing the sum total value of the set of eleven selected crops by the corresponding cropped area.

II. METHODOLOGY

In order to evaluate the statistical significance of change, the mean district values of the selected structural variables are calculated for the triennium 1959-61 and 1969-71, whereby the statistical significance of these mean values is tested with the help of the *t*-test. Then using the inter-district information the production functions are fitted for 1961 and 1971 and thereafter they are pooled together.

III. DISCUSSION OF RESULTS

The net state domestic product of Punjab at constant prices (1960-61=100) increased from Rs. 4410.7 million in 1960-61 to Rs. 7044.0 million in 1970-71, registering an annual growth rate of 4.8 per cent. During the same period, the contribution of animal husbandary to net state domestic product increased from 7.30 to 9.38 per cent. The share of agriculture in net state domestic product of Punjab increased marginally from 44.51 per cent in 1960-61 to 45.33 per cent in 1970-71 (Table 8.1).

Taking the primary sector alone, the growth rate of this sector during the period 1961-71 has been higher (5.2%) than that of the net state domestic products of Punjab (4.8%) in that period.

It becomes obvious from Table 8.2 that although the proportion of cultivators in the total male workers decreased from 46.3 per cent in 1961 to 42.8 per cent in 1971, yet the total number of workers (cultivators +agricultural labourers) increased from 55.9 per cent in 1961 to 62.8 per cent in 1971. The same is true for the whole of India (Table 8.2).

TABLE 8.1 : Growth and Structure of Agricultural Sector in Net State Domestic Product (Punjab during 1960-61 and 1970- 71)
Net State Domestic Product and Factor Cost and Constant 1960-61 Prices

(Rs. Crores)

		1960-61	*1970-71*	*Growth rate*
1.	*Total*	441.07	704.40	4.8
2.	Agricultural (Primary) Sector	229.40 (52.01)	386.96 (54.93)	5.2
	(a) Agriculture	196.30 (44.51)	319.33 (45.33)	5.0
	(b) Animal Husbandry	32.20 (7.30)	66.04 (9.38)	7.4
	(c) Forestry	0.80 (0.18)	1.45 (0.20)	1.5
	(d) Fishery	0.10 (0.02)	0.14 (0.02)	0.2

Note : Figures in brackets are percentage share of agricultural sector (with its sub-total in Net State Domestic Product.

Source : *Agricultural Census*, Punjab State, 1970-71.

TABLE 8.2 : Workers Engaged in Agriculture 1961 and 1971 : Punjab and All India

	Punjab (Figures in '000')		*India (Figures in '000')*	
	1961	*1971*	*1961*	*1971*
Cultivators	1606 (46.3)	1669 (42.8)	64,374 (61.6)	67,417 (56.0)
Agricultural Labourers	334 (9.6)	783 (20.0)	15,986 (15.3)	30,359 (25.2)
Other Workers	1528 (44.1)	1452 (37.2)	24,081 (23.1)	22,632 (18.8)
Total Workers	3,469 (100.0)	3,904 (100.0)	104,441 (100.0)	102.408 (100.0)
Population	11,135	13,551	123,357	226,219

Note : Figures in parentheses is percentage to total male workers.

Source : Agricultural Census, Punjab State 1970-71.

The mean district productivity per hectare increased from Rs. 969 in 1961 to Rs. 1607 in 1971; registering an annual growth rate of 5.2 per cent. Though the growth rate in the productivity per male workers is statistically significant (not the *t*-value) yet the growth rate of mean district productivity per male worker is less (3.6%) than that of the growth rate of productivity per hectare (5.2%). This can be explained by the increase in the number of agricultural labour force in Punjab as is evident from Table 8.2. The productivity per hectare of foodgrain crops[3] increased from Rs. 912 in 1961 to Rs. 1637 in 1971, thus recording an annual growth rate of 6.0 per cent. However, the productivity of non-foodgrain crops[4] did not increase significantly (note the t-value given in Table 8.3) during the same period.

The question which immediately strikes one's mind is : what has caused an increase in the productivity per hectare? Another related question is whether the increase in productivity is universal or confined only to a few crops. We have calculated the productivity per hectare of eleven crops which accounted for 83.4 per cent of the net sown area in 1971. An examination of the statistics given in Table 8.3 indicates that there has been an increase in the productivity per hectare of almost all the crops. It is to be noted that the increase in the productivity of foodgrain crops, especially wheat, maize and bajra has been greater than that of the non-foodgrain crops. In the case of wheat, maize and bajra, *t*-value are significant at one per cent level. It will be interesting to determine the factors which has caused such an increase in the productivity per hectare of these crops. This seems to have been made possible by the bio-irrigo-chemical technological breakthrough that took place in the mid-sixties. An examination of the statistics of the resource structure of Punjab's agriculture given in Table 8.4 should throw some light on it. The traditional capital goods represented by draft animals, ploughs and carts have lost their importance in Punjab's agriculture. This is confirmed by the negative growth rate of the mean district values of all the traditional farm capital resources (Table 8.4). On the other hand, the importance of modern farm capital goods like tractors, fertilizers, biological capital resource (HYV), human capital and pump-sets has increased in the resource structure of Punjab's agriculture. There has been a spectacular increase in the number of tractors, pump-sets and fertilizer consumption during the period 1961-71 (note the annual growth rate of mean district values of modern farm capital resources given in Table 8.4). The *t*-values of the variables representing modern farm capital resources especially mechanical traction power (number of tractors-units per lakh cropped hectares) and mechanical power for water lifting (number of pump-set unit per 100 gross irrigated area) are significant at one per cent level. The variable representing human capital is also significant at 5 per cent level. The proportion of area under HYV of food crops to total area under foodgrain crops also increased from nil in 1961 to 48.34 per cent in 1971.

Thus, an analysis of the statistics of modern capital resources given in Table 8.4 confirms that a substantial increase in the productivity per hectare of foodgrain crops has been made possibly by the bio-irrigo-technological breakthrough brought about by the scientists and the responsive farm community.

As regards the land resource structure, the cropped area has not increased significantly during the period 1961-71. The size of the cultivating units has declined significantly (for this the *t*-value for this period as given in Table 8.4, should be noted). The proportion of holding units has increased significantly over this period (again note that the t-values are significant at one per cent level). This implies that the number of small holdings has increased significantly during the period 1961-71.

Coming to the labour resource input, it becomes obvious from the figures given in Table 8.4 that the percentage of literate workers amongst rural male workers has increased significantly (note the *t*-value in Table 8.4). The rural labour force comprises cultivators, agricultural landless labourers and the workers. The mean district estimate of male cultivators per 100 cropped area has declined marginally from 33.64 in 1961 to 30.74 in 1971, registering negative growth rate of 0.9 per cent per annum. On the other hand, the mean district estimate of male agricultural labourers per '00' cropped hectares has increased significantly from 6.42 in 1961 to 13.83 in 1971, thus showing a growth rate of 8 per cent per annum. However, the inter-district variation in the agricultural labour population declined from 23.99 to 19.81 per cent. This accounts for the inter-district migration of labourers from low-wage districts to high-wage districts. The following factors seem to have contributed to the significant increase in the number of agricultural labourers in Punjab: (i) natural increase in the population of the labourers, (ii) shift of artisan and other workers from their traditional occupation to join the ranks of the agricultural workers, (iii) influx of the dis-possessed small farmers.

TABLE 8.3 : Gross Value of Output per Worker and per Hectare and Crop-wise Gross Value of Output per Hectare and Wage Rate During 1959-62 and 1969-72 Trienniums: Punjab

Sl. No.	*Description*	*Mean District Value*		*t-test for mean values*	*Annual growth rate of mean values*
		1961	*1971*		
1	2	3	4	5	6
1.	Gross Value of output of the selected eleven crop set (Rs. Crores)	36.88 (43.05)	71.46 (40.47)	3.19**	6.9
2.	Productivity per male worker (Rs.)	2094 (20.98)	2992 (24.13)	3.24**	3.6
3.	Productivity per Hectare (Rs.)	969 (14.50)	1607 (14.37)	7.18**	5.2
4.	Foodgrain crop Productivity per Hectare (Rs.)	912 (12.02)	1637 (18.50)	8.69**	6.0
5.	Non-foodgrain crop Productivity per Hectare (Rs.)	1400 (43.80)	1663 (35.75)	0.95	1.7
6.	Crop wise Gross Value of Produce per hectare				
(A)	Foodgrain				
	(i) Wheat	993 (18.95)	1930 (21.06)	6.34**	6.9
	(ii) Rice	1482 (24.16)	1784 (9.47)	2.39*	1.9
	(iii) Maize	621 (20.27)	840 (17.09)	3.52**	3
	(iv) Gram	787 (11.47)	913 (25.52)	1.22	1.5
	(v) Bajra	362 (31.16)	649 (19.41)	5.09**	6.0
	(vi) Barely	488 (16.99)	651 (27.87)	2.43*	2.9
(B)	Non-foodgrain				
	(i) Sugarcane	2691 (10.70)	3285 (12.85)	3.60**	2.1
	(ii) Potatoes	8027 (39.19)	6942 (53.30)	–0.69	-1.5
	(iii) Groundnut	1121 (18.62)	1353 (18.57)	2.30*	1.9
	(iv) Rape & Mustard	808 (22.69)	386 (18.57)	0.98	0.9
	(v) Cotton	471 (17.62)	561 (16.56)	2.22*	1.7
7.	Average wage rate to Agricultural Labourers per day (Man) at current prices (Rs.)	2.55 (17.25)	6.48 (11.30)	13.96**	9.7

Note : Figure in brackets are respective values of co-efficient of variation.

** Figures are significant at 1% level of significance.

* Figures are significant at 5 per cent level of significance.

(Critical value of t at 5% = 2.07 (or 2.09) at 1% = 2.82 or (2.86) with 22 degrees of freedom.)

TABLE 8.4 : Selected Set of Variables Depicting Changes in the Resource Structure of Punjab's Agriculture During 1961 and 1971

Sl. No.	*Description*	*Mean District Value*		*t-test for mean values*	*Annual growth rate of mean values*
		1961	*1971*		
1	2	3	4	5	6
(A)	**Traditional Farm Capital Resources**				
(i)	Animal Traction (Draft) Power Pair Units of Draft Animal 6 (working male cattle) per 100 cropped hectares	19.06 (32.21)	16.94 (36.36)	–0.77	–1.2
(ii)	Animal drawn Implements : No. of Plough units per 10 cropped hectares	2.67 (25.84)	2.63 (20.91)	–0.14	–0.2
(iii)	No. of Sugarcane crushers by worked bullocks per 100 cropped hectares	2.30 (83.48)	1.86 (79.03)	–0.58	–2.0
(iv)	No. of carts per 100 cropped hectares	7.00 (37.43)	5.50 (31.27)	-1.52	-2.4
(B)	**Modern Farm Capital Resources:**				
(i)	Mechanical traction Power : Tractor unit per lakh cropped hectares	87.85 (44.77)	698.29 (33.64)	8.11**	+23
(ii)	Chemical Capital Resources : Fertilizer Consumption (N+P+K) per cropped hectare	.0024 (62.5)	0.274 (34.31)	9.16**	108.1
(iii)	Biological Capital Resources : Proportion of Area Under HYV of food crops to total area under foodgrain crops	—	48.34 (29.66)	—	—
	Water Resource (Extent of Irrigation)				
(i)	Percentage of Irrigated area to net sown area	52.77 (33.05)	67.76 (31.12)	1.68	2.5
(ii)	Mechanical Power for water lifting : No. of pumpset units per 100 gross irrigated hectares	0.57 (61.40)	9.44 (42.26)	7.01**	32.4
(C)	**Labour Resources :**				
(i)	Male Agricultural Labourers per 100 cropped hectares	6.42 (23.99)	13.83 (19.31)	7.46**	8
(ii)	Male labour force per 100 cropped hectares	33.64 (19.68)	30.74 (16.14)	–1.11	–0.9

1	2	3	4	5	6
(iii)	Male Cultivators per100 cropped hectares	41.27 (19.43)	44.59 (14.89)	1.01	0.8
	Human Capital				
(iv)	Percentage of literate workers amongst rural workers—Human Capital	28.77 (19.22)	35.59 (20.78)	2.33*	2.1
(D)	**Land Resources :**				
(i)	Scale of operation :				
	(a) Gross cropped area (1000 hectares)	437 (54.92)	520 (50.58)	0.74	1.7
	(b) Cropped area under the set of eleven selected crops (1000 hectares)	401.8 (52.85)	452.9 (45.13)	0.54	1.2
	(c) Size of cultivating units (hectares)	5.03 (26.24)	3.06 (42.16)	-3.38**	-5.1
(ii)	Proportion of Holding Units to total Holding units :				
	(a) Proportion of holding units below 5 hectares total holding units (%)	59.66 (20.78)	82.65 (9.97)	4.89	3.3
	(b) Proportion of holding units below 1 hectare to total holding units (%)	7.34 (70.84)	37.51 (29.49)	8.45**	17.7

Sources :

1. Statistical Abstract, Punjab (various years)
2. Agricultural Marketing Board Chandigarh (For 1969-70 and 71 wholesale prices)
3. Director of Agriculture, Punjab (office) (for district-wise fertilizer consumption)
4. Economic and Statistical Organisation, Government of Punjab, Chandigarh (official documents)

Note : Figures in brackets are respective values of co-efficient of variation.

** Significant at 1 per cent level.

* Significant at 5 per cent level.

Critical Value of t at 5% = 2.07 or (2.00) at 1% = 2.82 (or 2.86) with 22 degrees of freedom.

Thus, a significant increase in the number of agricultural labourers confirms the belief that the process of proletarianisation is growing apace in the rural areas of Punjab.

Using the inter-district information, the Cobb-Douglas production functions are fitted for 1961 and 1971 and then those are pooled together.

Model

$$\text{Log } y = a_1 + b_1 \log x_1 + b_2 \log x_2 + b_3 \log x_3 + b_4 \log x_4 + b_5 \log x_5 + b_6 \log x_6 + b_7 \log x_7.$$

where

Y = Gross Value output of the selected eleven crop set at 1969-71 prices.

X_1 = Male labour force per '00' cropped hectare.

X_2 = Cropped area under the eleven crop set.

X_3 = Fertilizer (N+P+K) consumption per cropped hectare.

X_4 = Tractor units per lakh hectares.

X_5 = Percentage of irrigated area to net sown area.

X_6 = Literacy ratio of rural male workers.

X_7 = Dummy.

In order to know whether there exists the problem of multi-collinearity the zero-order correlation matrix has been computed for all the variables. Since the variable—'draft animal pair unit' per 100 cropped hectares was highly correlated with another independent variable namely cropped area under the eleven crop set (r = –0.700), the variable representing 'draft animal' was dropped. Similarly, because of the high correlation between the fertilizer consumption per cropped hectare and the pumpset units per 100 gross cropped hectares (r = 0.914), the latter variable was dropped. (see Table 8.5 for the correlation of co-efficient among the independent variables).

Thus having dropped the two independent variables (draft animal and pump-set unit), and then using the inter-district information for the rest of the variables, the Cobb-Douglas production function is fitted for 1961 and 1971. Further, the time-series-cum-cross-section are pooled together. The results of the production function are given in Table 8.6. For 1961, the significant variables are the male labour force, gross cropped area under the eleven crop-set the fertilizers consumption, and tractor units. In other words, these four variables have significant impact on the gross value output of the selected eleven crops. For the year, 1971, the only variable significant is the cropped area under the eleven-crop set. Under the pooled model the variables namely gross cropped area, fertilizer consumption and irrigation come out to be significant impact on the gross value of the selected eleven crops.

TABLE 8.5 : Zero-Order Correlation Matrix

	X_1	X_2	X_3	X_4	X_5	X_6	X_7	X_8	X_9	Y
X_1	1.000	–0.523	0.306	0.121	0.704	0.491	–0.064	0.664	0.283	–0.327
X_2		1.000	0.003	0.185	–0.700	0.179	0.303	–0.560	0.067	0.893
X_3			1.000	0.934	–0.066	0.914	0.283	0.565	0.980	0.408
X_4				1.000	–0.216	0.774	0.363	0.405	0.929	0.540
X_5					1.000	0.224	–0.320	0.546	–0.144	–0.641
X_6						1.000	0.024	0.733	0.877	0.199
X_7							1.000	–0.289	0.286	0.501
X_8								1.000	0.472	–0.295
X_9									1.000	0.445
X_{10}										1.000

X_1 = Male labour force per '00' cropped hectare (Nos.)
X_2 = Cropped area under the eleven crop set.
X_3 = Fertilizer (N+P+K) per cropped hectare.
X_4 = Tractor unit per lakh hactare.
X_5 = Draft animals pair units per '00' cropped hectares.
X_6 = Pump set units per '00' gross cropped hectares.
X_7 = % of irrigated area to net sown area.
X_8 = Literacy ratio of rural male worker.
X_9 = Concentration ratio of cultivating units.
Y = Gross value output of the eleven crops
Crop set at constant (1968–71) prices

TABLE 8.6 : Estimates of Agricultural Production Function on Inter-district Data of Punjab for 1959-62 and 1969-72

	Constant	X_1	X_2	X_3	X_4	X_5	X_6	X_7	X_8	R^2
1961	4.916*	1.708*	0.752*	0.510*	0.303*	–0.129	–2.084			0.994
	(1.185)	(0.584)	(0.077)	(0.174)	(0.134)	(0.105)	(0.857)			
1971	3.332	–0.073	1.064*	0.363	–0.141	0.184	0.060			0.995
	(0.987)	(0.190)	(0.082)	(0.260)	(0.118)	(0.193)	(0.324)			
Pooled	3.397	0.127	0.952	0.199*	–0.041	0.174*	–0.117	–0.195		0.983
	(0.606)	(0.210)	(0.057)	(0.074)	(0.072)	(0.080)	(0.280)	(0.142)		

X_1 = Male labour force per '00' cropped hectare.
X_2 = Cropped area under the eleven crop set.
X_3 = Fertilizer (N+P+K) per cropped hectare.
X_4 = Tractor units per lakh hectare.
X_5 = Percentage of irrigated area to net sown area.
X_6 = Literacy ratio of rural male workers.

Note : Figures in parenthesis are values of respective Standard Errors
*Significant at five per cent level of significance

IV. CONCLUSIONS

During the period the share of agriculture, forestry and fishery in the net state domestic product of Punjab, remained almost the same, but the share of animal husbandry in the net state domestic product increased from 7.30 per cent in 1961 to 9.38 per cent in 1971.

There has been a significant increase in the number of leadless agriculture labourers in Punjab during the period 1961-1971. This analysis confirms the belief that the process of proletarianisaiton is growing apace in the rural Punjab. The size of the cultivating units has also declined significantly. The traditional capital goods (draft animals, ploughs and carts) have lost importance while the use of modern capital goods (tractors, pump-sets, fertilizers, human capital, high-yielding varieties) has increased significantly.

The pooled time series-cum-cross-section estimates of the agriculture production on the inter-district data on Punjab for 1961 and 1971 also indicates that the modern capital resource variables such as mechanical traction power (proximated by tractor units per lakh cropped hectare), chemical capital resource (proximate by fertilizer consumption per cropped hectare), extent of irrigation (represented by percentage of irrigated area to net sown area) did have significant impact on the gross value output of most of the selected crops.

Note : Reproduced from *Productivity,* January-March, 1981 with kind permission of the publisher.

NOTES

1. This work is an adoption of a similar exercise done for Gujarat State, India, For details see Singh, B., *Changing Structure of Agriculture in Gujarat,* Sardar Patel Institute of Economic and Social Research, Reprint Series 17, earlier published in Lakdawala, D.T. & Mody, R.J. (ed.) Gujarat Economic Association, Conference Papers, October, 1977.
2. Worked by bullocks only.
3. Foodgrain crops include maize, rice, gram, bajra and barley.
4. Non-foodgrain crops include sugarcane, potatoes, groundnut, rape, mustard and cotton.

REFERENCES

Bhalla, G.S. (1973) : *Changing Structure of Agriculture in Haryana.*

Johnson, Bruce, F. (1970) : Agriculture and Structural Transformation in Developing Countries : A Survey of Research, *The Journal of Economic Literature,* Vol. VIII, No. 2, June, 1970, pp. 369-404.

Saini, G.R. (1971) Holding Size, Productivity and Some Related Aspects of Indian Agriculture, *Economic and Political Weekly,* June 1971.

Vyas, V.S. (1979) : Some Aspects of Structural Change in *Indian Agricultural Economics* Vol. xxxii, No. 1, January-March 1979, pp.1-18.

Vyas, V.S. (1979) : Structural Change in Agriculture and the Small Farm Sector: *Economic and Political Weekly,* Vol. XI Nos. 1 & 2, January 10, 1976, pp. 24-33.

9

RELATIVE LINKAGES OF INDIAN AGRICULTURE

S. SURYAPRAKASH, M.V. SRINIVASA GOWDA &
J.V. VENKATARAM

Introduction

The history of economic development in the world has shown that most of the less developed countries are characterised by the dominance of primary sectors, especially agriculture, in the economic activity. Agriculture in these countries tends to be primitive, self-contained and subsistent. Most of the inputs required will be met from the farm by way of home grown seed, organic manure from owned livestock and family labour. Again, most part of the output produced on the farm is retained for final demand by farm household and very little goes for final demand out side the farm sector.

As the economies develop, there is going to be a gradual rise in the importance of secondary and tertiary sectors which start contributing, in sizeable measure, to the national income and employment generation. But this process will not be independent of the growth of agriculture sector. This is mainly because the core problems of the developing countries have their roots in rural areas which are inhabited by the vast majority of the population. Agriculture needs to provide sufficient low-priced food, raw material and manpower to the expanding industrial sector. Again, agriculture and rural population provides demand for non-farm inputs and wage goods (Datt and Sundaram, 1993).

Given such sectorol changes in the process of dynamic growth, it is envisaged that agriculture's dependence on itself declines and that on the other sectors of the economy increases as the nation's economy develops over time. It will be of importance to study this process over time to examine whether the agriculture sector of an economy is independent or inter-twined with the rest of the economy. This could be measured through the relative linkage.

Concept and Methodology

The sector-wide and economy-wide linkages of different sectors of an economy are helpful in analysing the production structure of an economy. This will help in arriving at suitable strategies for economic development. But these are not helpful directly in sector planning, more so, when the sector remains isolated from the rest of the economic system. In such cases, relative linkage is a more useful measure.

It should be noted that a sector in an economy will have links not only with the other sectors but also with itself. A sector's link with itself is intra-sector linkage. The sector's link with the other sectors or the economy is measured by sector-wide or economy-wide linkage. The former linkage *vis-a-vis* the latter linkage presents a picture of relative dependence (importance) of the sector. This relative linkage is useful in sector planning. Pal (1988) has given the operational significance of the relative linkage (RL).

Every sector of an economy can be divided into several sub-sectors. Each of these sub-sectors will have links with its parent sector on the one hand and the rest of the economy on the other. It may so happen that a sub-sector's linkage with the rest of the economy may be weak. But this does not mean that its links with the parent sector will also be weak. This information is reflected in the relative linkage.

The relative linkage is conceptualised and measured in the input-output framework. A static open input-output system of n-sectors is characterised by the transaction matrix $X = (X_{ij})$ and gross output vector $x = (x_j)$. If there exists a sector, say I, comprising nI (sub) sectors from the system, then the relative linkage (*RL*) for the sub-sector, say I, within sector *I* is defined as the ratio of its intra-sector linkage to economy-wide linkage.

The relative linkage can be worked out both on the input-use side and output-distribution side. The input *RL* of sub-sector *i* is given by

$$\text{Input } RL_i = \frac{\sum_{j \in I}^{nI} X_{ji}/x_i}{\sum_{j=I}^{n} X_{ji}/x_i} \qquad \ldots (1)$$

The input RL of a sub-sector is the ratio of the amount of input drawn from its parent sector (intra-sector linkage) to the input drawn from the entire system (economy-wide linkage) to produce a unit of its output. This ratio indicates the relative dependence of the sub-sector on the system.

The output RL of sub-sector *i* is given by

$$\text{Output RL}_i = \frac{\sum_{j\in I}^{nI} X_{ij}/x_i}{\sum_{j=I}^{n} X_{ij}/x_i} \qquad \ldots (2)$$

The output RL of a sub-sector is thus the ratio of the proportion of its total output distributed to its parent sector to the proportion of its output distributed to the entire system as input. The output RL presents a picture of the relative importance of the sub-sector.

The relative linkage ranges between 0 and 1. A value of 0 on input-use side implies that the sub-sector is completely dependent on the system with no intra-sector dependence. A value of *1* indicates that the sub-sector is entirely dependent on its parent sector for all its input requirement. The ranges on the output-distribution side could similarly be interpreted. A higher magnitude of relative linkage (i.e., closer to 1) implies that the sub-sector in question is independent of the system. It draws most of its input only from the parent sector and the intermediate demand for its output is mostly from the parent sector. In such a case, planning done for the whole economy for its development will have no effect on the sub-sector and it calls for sectoral planning.

Data

In the light of the foregoing, an attempt was made to study the relative linkages of Indian agriculture. Attempt was made to test the notion that Indian agriculture continues to be a subsistence sector using primary inputs and producing output mainly for final consumption. For the purpose of the present study, all the four 60-sector input-output tables prepared by the Central Statistical Organization, Government of India, New Delhi were used. These Tables 9.1, 9.2 and 9.3 pertained to the years 1968-69, 1973-74, 1978-79 and 1983-84. All the entries in these tables were expressed at constant (1970-71) prices using the wholesale price indices of different years. The sectors of food crops, cash crops, plantation crops and other crops were clubbed to form the agriculture sector. The relative linkages of these four sectors were worked out considering them as sub-sectors of agriculture sector. Food crops comprised paddy, wheat, jowar, bajra, maize, gram and pulses, while the crops of sugarcane, tobacco, groundnut, jute and cotton constituted cash corps. Plantation crops include coffee, tea, rubber and coconut. All other cereals, oilseeds, fruits and vegetables constitute 'other' crops.

Findings

As indicated earlier, the relative linkages were attempted both on thc input (backward) side and on the output (forward) side.

A. *Relative Linkages : Input-use side*

(i) Food crops : The relative linkage of food crops sub-sector had declined over the years from 0.4620 to 0.3405 (Table 9.1). The intra-sector linkage (the linkage with agriculture sector) was almost of the same magnitude over time, except for a decline in 1973-74. On the other hand, the economy-wide linkage strengthened over the years, barring a slight fall in 1973-74.

Thus, the fall in relative linkage is mainly contributed by the increase in the economy-wide linkage rather than by the fall in intra-sector linkage. This shows that this sub-sector has opened itself up more to the economy for its inputs. But, this has not resulted in a noticeable reduction in its inwardness. From the composition of the sector, one can observe that crops like paddy and wheat, with their increased dependence on off-farm inputs, have contributed to the increased openness of the sector. Crops like bajra, gram and pluses—in whose case the availability and adoption of improved technology were poor during the period under reference have thwarted the decline in intra-sector linkages caused by paddy and wheat.

(ii) Cash crops : The relative linkage in the case of cash crops had fallen drastically over the years, from 0.4198 in 1968-69 to 0.2757 in 1983-84 (Table 9.1). The causes for this fall were increased economy-wide linkage and decreased intra-sector linkage. But the contribution of the former was more pronounced than the latter.

Continued production of jute and tobacco mostly with traditional technology could be the cause for the negligible fall in the sector's intra-sector linkage.

(iii) Plantation crops : The relative linkage of this group was nil during 1968-69 and 1973-74 (Table 9.1). Later on it became positive, but declined in 1983-84. This phenomenon was caused by the presence of intra-sector linkage in 1978-79 and 1983-84, though the economy-

TABLE 9.1 : Input-use side relative linakges in Indian agriculture

Sub-sector	*Intra-sector linkage*	*Economy-wide linkage*	*Relative linkage*
I. Food crops			
1968-69	0.1146	0.2482	0.4620
1973-74	0.0953	0.2176	0.4378
1978-79	0.1264	0.3099	0.4077
1983-84	0.1113	0.3268	0.3405
II. Cash crops			
1968-69	0.0617	0.1469	0.4198
1973-74	0.0506	0.1567	0.3226
1978-79	0.0502	0.1743	0.2882
1983-84	0.0548	0.1987	0.2757
III. Plantation crops			
1968-69	0.0000	0.0417	0.0000
1973-74	0.0000	0.1034	0.0000
1978-79	0.0069	0.1493	0.0462
1983-84	0.0034	0.1702	0.0199
IV Other crops			
1968-69	0.0173	0.0882	0.1964
1973-74	0.0262	0.0897	0.2920
1978-79	0.0157	0.0977	0.1601
1983-84	0.0222	0.1034	0.2142

Notes : 1. Intra-sector linkage depicts the linkage of the sub-sector with the agriculture sector.

2. Economy-wide linkage depicts the linkage of the sub-sector with the entire economy.

wide linkages strengthened over the years. Generally, the traditional plantation crops like coffee and tea have little to do with the other crop sectors and hence there is absence of intra-sector linkage in their case. Increased shares of coconut in plantation crops might be the cause for the presence of intra-sector linkage.

(iv) Other crops : In this case the relative linkage had fluctuated

without a definite pattern, but a marginal increase was observed in 1983-84 (Table 9.1). The components of the relative linkage indicate that the economy-wide linkage had grown over years, although marginally. On the contrary, the intra-sector linkage fluctuated and ultimately rose by 1983-84.Thus, the movements in the relative linkage were mostly caused by the intra-sector linkage changes over the years.

Some of the crops in this group like castor, niger, sesamum, linseed, etc., were being cultivated mostly under rainfed conditions and on poor lands. The technological breakthrough was also not substantial for these crops. Under such situations, input use in general and use of off-farm (purchased) inputs in particular, would be tardy and depends upon normality or otherwise of the rainfall and the trend in output prices. This can explain the movement of relative linkage index of other crops over the years. Favourable prices, regular rainfall and improved technology would be able to improve the situation.

The overall situation considering the four sectors together, can be depicted as under, With minor variations, it can be generalised that the agriculture sector in India has shown a tendency to rely less on itself for inputs (in a relative sense). This is in conformity with the general pattern observed world-wide that with development, sectors open up more and increase their dependence on the economy. But this may not necessarily be a welcome feature as the present thinking is more towards organic farming and sustainable agriculture which clearly imply reductions in off-farm input use and an increase in the dependence on own inputs.

Another feature that can be observed is that the economy-wide linkages of a sub-sector changed over time partly due to the changes in growth rates of the sub-sector's output (Table 9.3). Thus, it may be said that the input-side relative linkages indirectly influenced by the output growth rate as well.

B. *Relative linkage : Output-distribution side*

Like on the input side, the relative linkages were also worked out on the output-distribution side also and the same are presented in Table 9.2. A notable feature is that the relative linkages on distribution-side too moved in the same direction as those on the input-side.

(i) Food crops : As expected, the relative linkage declined over the years for the food crops sub-sector except for a marginal rise in 1978-79.

However, this decline was not substantial. Looking at the inter-temporal movements of the components of the relative linkage, one can observe that both the intra-sector and economy-wide linkages move in harmony with each other over the years. That is to say, the inter-temporal movements of the relative linkage were contributed by both the components.

The aforesaid situation cannot be explained through the growth rates in the sub-sector output (Table 9.3), since the output growth rates and

TABLE 9.2 : Output-distribution side relative linakges in Indian agriculture

Sub-sector	*Intra-sector linkage*	*Economy-wide linkage*	*Relative linkage*
I. Food crops			
1968-69	0.1093	0.1565	0.6985
1973-74	0.0917	0.1409	0.6505
1978-79	0.1244	0.1840	0.6761
1983-84	0.1075	0.1758	0.6116
II. Cash crops			
1968-69	0.0605	0.5482	0.1103
1973-74	0.0515	0.6844	0.0752
1978-79	0.0506	0.7467	0.0678
1983-84	0.0557	0.7831	0.0712
III. Plantation crops			
1968-69	0.0000	0.4045	0.0000
1973-74	0.0000	0.3512	0.0000
1978-79	0.0068	0.4735	0.0144
1983-84	0.0034	0.4949	0.0068
IV Other crops			
1968-69	0.0192	0.3925	0.0490
1973-74	0.0273	0.4132	0.0660
1978-79	0.0175	0.4103	0.0427
1983-84	0.0237	0.4861	0.0487

Notes : 1. Intra-sector linkage depicts the linkage of the sub-sector with the agriculture sector.

2. Economy-wide linkage depicts the linkage of the sub-sector with the entire economy.

relative linkages (as also their components) had not moved in a similar direction like in the case of input-side relative linkages. Food crops are mainly used for final consumption in a country like India. Again, the experience shows that output growth had been divergent because of a host of factors like unpredictable weather, changing relative advantages, poor technology adoption, etc. Final demand being large and increasing with increases in population and incomes, a negligible portion of food crop output went for intermediate demand during the period under study. A further probe into the situation for recent periods not covered by the present study may throw more light on the behaviour of relative linkage.

(ii) Cash crops : In this case the relative linkage has been declining over the years, except for a small rise in the last year under reference.

TABLE 9.3 : Compound growth rates of sectoral outputs in Indian agriculture

(Percentage)

Period	*Food crops*	*Cash crops*	*Plantation crops*	*Other crops*
1968-73	4.53	5.06	7.52	1.02
1973-78	1.50	–1.44	–3.73	3.11
1978-83	2.40	3.52	–0.17	2.81
1968-83	2.80	2.34	1.10	2.31

Note : The growth rates were computed using the formula :

$$g = e^{\ln \frac{Y_T/Y_o}{T}} - 1$$

where

g = annual compound growth rate
Y_T = output in year T
Yo = output in base year
T = number of years

The deviation in 1983-84 was mainly caused by the increase in intra-sector linkage since the economy-wide linkage has been rising steadily. Thus, it is the economy-wide linkage which has been contributing to the changes in relative linkage over the years. This trend is easy to comprehend since the cash crop outputs, by their very nature, mostly go for intermediate use. Again, the influence of economy-wide linkage on relative linkage continued despite erratic growth in the output of the sub-sector. The pattern in this sub-sector was as expected.

(iii) Plantation crops : The relative linkage (RL) of plantation crops on the distribution side moved quite similar to the RL on input side. The RL was zero upto 1973-74, turned positive later on but declined by 1983-84. The changes in RL were mostly accounted for by the changes in the intra-sector linkage despite the increase in the economy-wide linkage.

As indicated while discussing the inputs RL of this sub-sector, the positive intra-sector linkage in the later part of the period under reference might be due to the coconut crop which is partly consumed by the sub-system. Such a thing does not happen with the other crops of the group.

(iv) Other crops : The relative linkage of this sub-sector exhibited a case of fluctuating RL. But all the three, viz., intra-sector, economy-wide and relative linkages changed in the same direction over time. This implies that both intra-sector and economy-wide linkages have contributed to the changes in relative linkage. The circumstances leading to such a phenomenon have been explained in the discussion on the input-side RL of this sub-sector.

Looking to the four sub-sectors, one can observe that, as far as the relative linkages on the distribution side are concerned, the situation has not changed much during the 15-year period under study. But, the little change that had taken place has been on anticipated lines.

Pal (1988) compared the relative linkages of the foodgrains sub-sector in India over the period 1953-54 to 1964-65. He observed that foodgrains had strong relative linkages (input as well as distribution) indicating a weak structural inter-dependence. These findings when compared with those of the present study show that things have changed after 1965 when new technology was introduced by way of green revolution.

Conclusion

Agriculture is largely traditional. But as the economy develops, agriculture too turns towards the market, becomes more and more independent from itself and integrates with the economy. The study of the relative linkages in Indian agriculture during 1968-83 has shown that food crops and commercial crops have opened up more to the economy for inputs while the other two crop sectors did not change much. As far as the output distribution side is concerned, the situation has not changed much. This implies that sector planning is essential for the development of Indian agriculture to bring about complete integration with the rest of the economy. This has been amply demonstrated in the case of Punjab and Haryana states.

REFERENCES

Bhalla, G.S., Chadha, G.K. Kashyap, S.P. and Sharma, R.K., (1990), *Agricultrual growth and structural changes in the Punjab economy: An input-output analysis.* (Report No. 82), International Food Policy Research Institute, Washington, D.C.

Carter, A.P. and Brody, A. (1970), *Applications of input-output analysis.* North Holland Publishing Company, Amsterdam.

Datt R. and Sundaram, K.P., (1993), *Indian economy.* S. Chand and Company Ltd., New Delhi.

Hou, Chi-Ming and Yu, Tzong-Shian (eds.), (1982), *Agricultural development in China, Japan and Korea.* Academica Sinica, Taipei, Tiwan, ROC.

Pal, D.P., (1988), *Structural Independence and Development.* Himalaya Publishing House, Bombay.

Rangarajan, C., (1982), *Agricultural Growth and Industrial Performance in India.* (Report No. 33) International Food Policy Research Institute, Washington, D.C.

10

ACHIEVING SUSTAINABILITY AND EXTENDING THE GREEN REVOLUTION

M.S. SWAMINATHAN

Introduction

"Intensive cultivation of land without conservation of soil fertility and soil structure would lead ultimately to the springing up of deserts. Irrigation without arrangements for drainage would result in soils getting alkaline or saline. Indiscriminate use of pesticides, fungicides and herbicides could cause adverse changes in biological balance as well as lead to an increase in the incidence of cancer and other diseases, through the toxic residues being present in the grains or other edible parts. Unscientific tapping of underground water would lead to the rapid exhaustion of this wonderful capital resource left to us through ages of natural farming. The rapid replacement of numerous locally adapted varieties with one or two high yielding strains in large contiguous areas would result in the spread of serious diseases capable of wiping out entire crops, as happened prior to the Irish potato famine of 1845 and the Bengal rice famine of 1942. This process has already begun during 1967 with ergot of *bajra* and may happen with reference to rusts of wheat and virus diseases of rice during 1968. Therefore, the initiation of exploitative agriculture without a proper understanding of the various consequences of every one of the changes introduced into traditional agriculture and without first building up a proper scientific and training base to sustain it, may only lead us into an era of agricultural disaster in the long run, rather to an era of agricultural prosperity."

Thus, it was clear from the beginning that the indiscriminate use of chemicals and non-adoption of ecologically sound crop-management practices would do harm in the long-term in sustaining high yields. How then can we foster an ecologically sound green revolution? Fortunately timely scientific efforts initiated by the Indian Council of Agricultural Research and Agricultural Universities made it possible to combine advances in production and productivity in wheat and rice with yield stability. The task now is the addition of the dimensions of ecological sustainability and equity in the green revolution pattern of agricultural advance.

Achieving Sustainability

A strategy for sustainable living titled *Caring for the Earth,* prepared by IUCN, UNEP and WWF defines sustainable development as "improving the quality of human life while living within the carrying capacity of supporting ecosystems." A dynamic concept of carrying capacity would imply in operational terms the conservation of natural ecosystems as well as their continuous improvement through research, training, technology, community cooperation and public policies. Sustainable Nutrition Security has been defined as "providing physical and economic access to balanced diets and safe drinking water to all people at all times" (Swaminathan, 1987). What kinds of re-orientation do we need in agricultural research and development strategies and methodologies to achieve this goal, keeping in mind the fact that the number of people affected now by hunger and malnutrition is of the following order?

Indicator	*Total population*
Energy insufficient for work	1.053 million
Energy insufficient for normal growth of children, minimal activity of adults	477 million

Source: Millman, 1991.

As we approach 1992 when international attention will be focussed on issues relating to sustainable development there will be increasing interests in all aspects of converting the concept of sustainability into an operational reality. Both in agriculture and industry the concept of sustainability has so far been considered predominantly at the level of economics. Farmers, for example, generally take decisions on the technologies to be adopted on the basis of cost, risk and return calculations. Social sustainability has also been an important factor since farmers well still prefer to grow crops which are needed for their home consumption, whether or not those crops are desirable from the standpoint of ecologically sound land use. Thus even today farmers in the higher altitudes of the Himalayas as well as Western and Eastern Ghats of India grow annual crops in highly erodable soil.

Economics-driven growth has led to a $20-trillion global economy. The decline in environmental quality has however underlined the need for harmonising the logic of economics with that of ecology. An important consequence of human action is the gradual conversion of the surface of the earth into wastelands and degraded lands, Globally 15 per cent of the total earth surface has undergone human induced soil degradation according to International Soil Resources Inventory Centre at Wageningen in the Netherlands. The study has also revealed that about 24 per cent of the human occupied territory of the earth is degraded by anthropogenic factors. At least 66 million ha of irrigated land is affected by secondary salinization. About 1 million ha, much of it prime farm land in rainfed areas, is being lost each year to urbanization. There are also similar figures with reference to water pollution and groundwater exhaustion. Compounding these problems is the gradual division of forest lands for a variety of other uses, thereby resulting in the loss of habitats rich in biological diversity. In India, crop land occupies 47 per cent of the geographical area. In USA this percentage is 21 (Figs. 10.1 and 10.2)

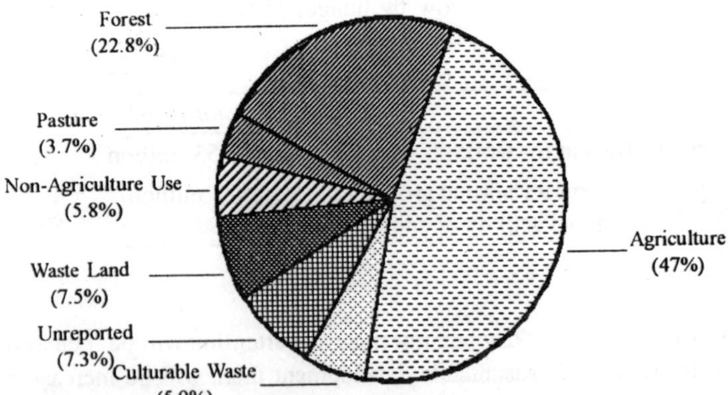

Fig.10.1 : Land Utilisation in India

Source : The State of Forest Report, 1987, Forest Survey of India, Ministry of Environment and Forest, Government of India

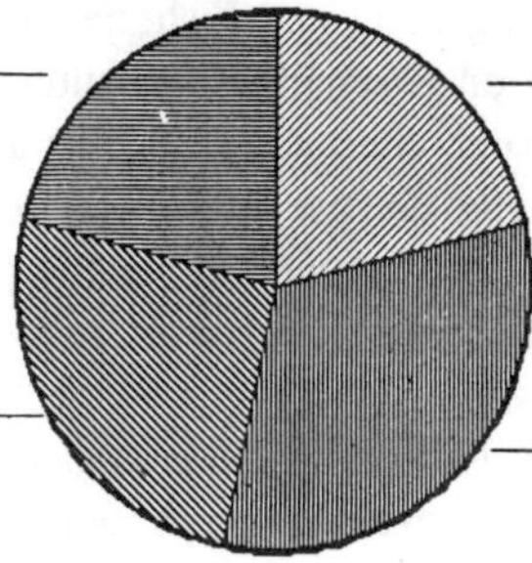

Fig. 10.2 : Land Utilisation in United States

Source : H.T. Fry and R.W. Hexem, Major Uses of Land in the United States; 1982 Agricultural Economic Report No. 535 U.S. Department of Agriculture, Economic Research Service; 1985.

New Paradigm of Development

It is in the content of the above macro-scenario that we have to consider methods of achieving continuous advances in biological productivity in an ecologically sustainable manner. The new paradigm of development now emerging places as much stress on ecological sustainability as on economic and social sustainability. Industrialised countries have intensified their efforts to arrest soil erosion, conserve water and biological diversity and reduce the consumption of mineral fertilizers and chemical pesticides and at the same time stabilise the per hectare productivity at the current high levels. The average yield per ha of cereals ranges from five to eight tonnes in most industrialised countries, with the exception of countries like Canada, where the average yields are lower due to the crops being mostly rainfed. The Food Security Act of the United States adopted in 1985 is a good example of political action for promoting conservation-based agriculture. The U.S. Food Security Act is also unique in the sense it addresses concurrently issues relating to conservation, production and consumption.

Most countries in Europe have also developed programmes to gradually reduce the use of pesticides. For Example, Scandinavian countries have enacted legislation to reduce pesticide use, with Denmark calling for a reduction of 50 per cent by 1997 and Sweden setting a target in 1988 of a 50 per cent cut back within only five years. Their lead has already been followed by countries in Europe (Ministry of Agriculture. Nature Management and Fisheries, the Netherlands 1990; Warrell, 1990. A report on alternative agriculture (National Research Council, 1989) defined such alternative systems as those incorporating natural processes, reducing the use of inputs from off-farm sources, making greater use of the biological and genetic potential of plant and animal species, *ensuring the long-term sustainability of current production levels* (italics mine), and improving farm management and the conservation of soil, water, energy and biological resources. Significant government support for the development of such systems exists; the US Food, Agriculture, Conservation and Trade Act, 1990 farm Bill), for example, authorises US $40 million annually for Low Input Sustainable Agriculture (LISA) research (Madden and O'Connell, 1990).

In industrialised countries the major concern is naturally the maintenance of the status quo in yield since this will be more than adequate to meet their home and export needs. In India, China and many other developing countries in contrast, the challenge is how to produce more from diminishing land and water resources. Therefore, such countries will have to develop appropriate technologies and public policies for achieving the desired rates of growth without detriment to the long term production potential of soil and water.

Impact on Employment

Another challenge which is somewhat unique to countries like India, Bangladesh, China and Indonesia is the need for generating more jobs in the farm sector. In spite of the diversification of their economies, most developing countries still depend upon the farm sector for providing most of the jobs. Although the share of agriculture in India's overall gross domestic product has come down from 59.6 per cent in 1950 to about 30 per cent now, over 75 per cent of rural population still depend on agriculture for their livelihood. The creation of new jobs for over 100 million persons by the year 2000 will call for a growth in employment of 4 per cent per annum as compared to the present rate of about 2.0 per cent. Out of this four per cent growth rate, nearly three per cent may have to come from crop and animal husbandry, fisheries, forestry and agro-industralised. This is not the kind of problem which the industrialised countries face. In other words, in addition to the requirements of economic efficiency and ecological sustainability, employment generation becomes an important consideration in technology development and dissemination. Those technology which may be desirable from the point of view of income but which may lead to the displacement of labour, will have adverse social consequences. This is particularly true for women-specific occupations. Farm mechanisation has often displaced women from their traditional jobs. So also is the substitution of organic manuring methods with mineral fertilisers. Women have been traditional seed selectors but large seed companies have tended to displace them from their role in *in situ* conservation of land races. The starting point of both technology and public policy development is subjecting the new techniques/policies to a *Sustainability Impact Analysis* using an illustrative methodology indicated below :

Technology/policy	*Impact*		
	Economics	*Ecology*	*Employment*
Mineral Fertiliser	Positive	Negative	Positive
Chemical Pesticide	Positive	Negative	Positive
Integrated Nutrient Supply	Positive	Positive	Positive
Integrated Pest Management	Positive	Positive	Positive
Heavy Machinery	Positive	Negative	Negative
High price incentives for water loving plants in semi-arid areas	Positive	Positive	Positive

The above is an illustration of the kind of analysis needed prior to the introduction of new technologies and public policies designed to accelerate agricultural growth. The answer to the different categories of questions will obviously vary depending on factors such as farming system and socio-economic and agro-ecological conditions.

Principles of Sustainability

Basic Principles

The following nine principles will have to be kept in view to achieve success in promoting ecological agriculture.

(i) Land

Based on both biological potential and biological diversity, land can be classified into conservation, restoration and sustainable intensification areas. Diversion of land suitable for sustainable intensification for non-farm uses should be prevented by legislation. Such land should be objected to a continuous soil health monitoring. Soils with diminished biological potential, also referred to often as waste or degraded lands, should be improved through the application of the principles of restoration ecology. Conservation areas rich in biological diversity must be protected in their pristine purity.

(ii) Water

Effectiveness in water saving, equity in water sharing and efficiency in water delivery and use are important for the sustainable management of available surface and groundwater resources. There should be an integrated policy for the conjunctive and appropriate use of river, rain, ground and sea water and for the recycling of sewage water and industrial effluents.

(iii) Energy

Integrated systems of energy management involving the use of renewable and non-renewable resources of energy in an appropriate manner are essential for achieving the desired yield levels.

(iv) Nutrient supply

Soils in India are often not only "thirsty" but also "hungry". Inputs are needed for output. Therefore, what we need is a reduction in the use of market purchased chemical inputs and not of input *per se.* It is in this context that integrated systems of nutrient supply assume importance. The components of the integrated nutrient supply system suitable for easy adoption include crop rotations, green manures and bio-fertilizers. Biodynamic system that make significant use of compost and humus will help to improve soil structure and fertility.

(v) Genetic diversity

Genetic diversity and location-specific varieties are essential for achieving sustainable advances in productivity. Traditional systems of farming depended heavily on *in situ* conservation of genetic variability in the form of numerous land races, mostly by women. Genetic homogeneity characteristic of modern agricultural systems only leads to greater genetic vulnerability to biotic stress. Participatory plant breeding efforts jointly with rural families will help to achieve a desirable blend of traditional and frontier technologies. Diversity of crops and varieties will be the result of such a joint endeavour, thus enhancing stability of yield.

(vi) Pest management

The control of weeds, insect pests and pathogens is one of the most challenging jobs in tropical and subtropical agriculture. Here again, an integrated pest management system needs widespread development and adoption. The conservation of natural enemies of pests is important for minimising the use of chemical pesticides and for avoiding the multiplication of insecticide-resistant pests. Botanical pesticides like those derived from neem need popularisation. The conservation and wise use of genetic diversity are essential for breeding strains possessing multiple resistance to biotic and abiotic stresses. Selective mircrobial pesticides offer particular promise, of which strains of *Bacillus thur ingiensis* (Bt) serve as an example. Transgenic techniques have made the transfer and expression of the Bt toxin possible in several crops. The applications of genetic engineering as applied to pest control are growing rapidly. It is however necessary to ensure that the "natural pesticides" synthesised by plants for their projection do not pose health hazards. Tests for wholesomeness should keep this aspect in view.

(vii) Post-harvest systems

Whole plant utilisation methods and the preparation of value added products from the available agricultural biomass are important both for enhancing income and for ensuring good nutritional and consumer acceptance properties. Both producers and consumers will not derive benefit from production advances if there is a mismatch between production and post-harvest technologies. Drying, storage and marketing techniques should be such that they not only do not make much demand on non-renewable sources of energy but they also prevent quantitative and qualitative damage to foodgrains or other agricultural commodities.

Mycotoxins and bacterials food infections often become a serious problem and hence greater attention to the qualitative aspects of food storage and distribution is urgently needed. Nucleic acid probes and monoclonal antibodies can help in rapid diagnosis, and hence more investment in research and training in this field will be worthwhile.

(viii) Systems approach

Ecological and economic sustainability is enhanced when the available land, water, energy and credit resources are utilised in a mutually reinforcing manner. A systems approach involving integrated attention to crop and livestock farming and to agro-forestry and aquaculture will be helpful in generating more jobs and income and at the same time protect soil health.

(ix) Location specific research and development

A fundamental requirement for sustainability is participatory research and training. This will call for new patterns of organisation of research, with scientists and farm becoming partners in the development and dissemination of new technologies. Only then can we promote and protect national and global food security within the limits of the carrying capacity of the supporting ecosystems. "Think nationally but plan and act locally" is the only relevant method of promoting sustainability in farming.

Under conditions of small and fragmented holdings, the adoption of the above principles of sustainability will be possible only if all farming families in a village or a watershed or the command area of an irrigation project cooperate. For example, although IPM techniques have been available in the case of rice and cotton pests for some years now, their field adoption has been slow due to inadequate attention to the promotion of group cooperation among small farm families. Social organisation is as important as technological innovation for achieving continuous improvements in biological productivity on an ecologically sustainable basis.

Sustainable Forestry

Since forest denudation is one of the most serious factors promoting the sustainability of both nutrition and food security, would like to indicate how in India forests can be conserved and utilised to meet the wood requirements of the community and small and large scale industries. Industrial needs of wood are growing. For example, the gap between demand and availability of raw material for the paper and pulp industry in India will grow from the present 6.3 lakh tonnes to 15.5 lakh tonnes in the year 2000 and to 46.5 lakh tonnes by the year 2015.

In addition, the Planning Commission of India has estimated the needs for fuelwood and fodder as follows:

(a) fuelwood demand would be about 306 million tonnes in 1991, 343 million tonnes in 1996 and 383 million tonnes in 2001; and
(b) fodder demand will be 635 million tonnes in 1991, 648 million tonnes in 1996 and 660 million tonnes in 2001.

The Forest Survey of India estimates that the maximum quantities of wood that the country can produce annually in reasonably ideal conditions in 90 million cu.m. Obviously the time has come when we should look at the triple need of conservation, community needs and commercial requirements with regard to wood in an integrated manner. One way of solving this problem is the initiation of a Sustainable District Forestry Programme (SDFP) in all the districts of India. I have used the prefix "sustainable" to stress that the programme should be ecological, economically and socially sustainable. The three major components of SDFP are:

Conservation Forestry

India is one of the 12 major mega diversity countries in the world because of its wide spectrum of soil and climatic conditions and its position at the Junction of African, Palaeartic and Oriental regions. Government Forest Departments control about 75 out of 320 million ha of the land area. Out of this only 30 million ha have good canopy. The good canopy forests have poor accessibility and this has saved them. Harvests for industrial purposes as well as for the needs of the people of adjoining areas have resulted in 45 million ha of forests being denuded to varying extents.

About 3 per cent of the land area is under nature reserves. Today, protecting such protected area itself is becoming a major challenge. The Commission on National Parks and Projected Areas of IUCN in its report on *Threatened Protected Areas of the World* published in November 1990 has listed the following protected areas in India as endangered.

(a) Manas Wild Life Sanctuary (A World Heritage site), Assam
(b) Kaziranga National Park, Assam
(c) Gulf of Kutch Marine National Park, Gujarat
(d) Keibal Lamjao National Park, Manipur
(e) Gir National Park, Gujarat

The following regions rich in biological wealth have also become "hot spot" areas from the point of view of genetic erosion.

(a) Agastyamalai Hills, Western Ghats
(b) The Silent Valley and the New Amarambalam Reserve, Western Ghats
(c) The Periyar Lake area, Western Ghats
(d) Eastern Himalayas
(e) Western Himalayas

It is high time, as suggested in the document *Caring for the Earth,* such areas are developed into Biodiversity Conservation Regions (BCRs) jointly by local communities, government agencies and corporate and other interests. This will help to promote *in situ* and *ex situ* conservation of both fauna and flora, including micro-fauna and micro-flora.

Community Forestry

The present social forestry programmes often tend to be Government forestry for the people, rather than people's own forestry for meeting their felt needs. The choice of tree species, their propagation, planting, protection and post-harvest utilisation should all be under community control either through village organisations or Tree Cooperatives or under any other acceptable and socio-culturally compatible method of management.

Commercial Forestry

Wood is needed for a wide variety of domestic and industrial needs. The Commercial Forestry programme should be based on the best available genetic material, ecological management methods and efficient post-harvest technology. The work done within India by several industries as well as non-governmental organisations show that yields from forest plantations can be increased from the present 40 AST/ha/7 years to about 150-200 AST/ha/7 years through the planting of superior clones and scientific management of soil health and plant growth.

I have often referred to "green revolution" as "land saving agriculture" because of the adoption of the productivity pathway in meeting the additional food requirements. Similarly, we need land-saving forestry from the point of view of meeting commercial and social requirements. For example in Karnataka, the annual productivity of the Government Eucalyptus plantation is as low as 1.5 to 2 tonnes per hectare per year. On the other hand, farmers who take to tree farming on private lands harvest 10 to 15 tonnes per hectare per year. Prof. Madhav Gadgil of the Indian Institute of Science has calculated that at an yields of 10 tonnes per hectare annually. India will need only 3 to 6 million hectares, i.e. about 2 per cent of all cultivated land, to meet the total commercial timber needs of less than 30 million tonnes per year. Such as land-saving forestry programme would eliminate the need for the harvesting of wood from the remaining 60 million hectares of reserve forest land.

The concept of sustainable yield should not be considered only with reference to the areas which have experienced the Green Revolution. It is equally relevant to other categories of farming characterised by low productivity at present. Nearly 65 per cent of crop land in India is still rainfed. There is a long coastal area of over 7500 km in addition to several islands. There are also extensive hill ecosystems including the Himalayas which stretches 2400 km in length and 300 km in width. The future of agriculture in the Indo-Gangetic plains, which is a major food basket region of India and the world, will be determined by the results of the efforts currently under way for caring for the Himalayas.

Achieving Sustainability

The problem facing us today is not so much one of discovering what must be done to ensure sustainability as learning how to do it. The technologies which can promote sustainability rely heavily on knowledge as a substitute for capital, farm grown inputs as substitutes for market purchased ones and community cooperation as a supplement to individual action. Location specific research will be needed to develop such a technological package. Concepts such as integrated nutrient supply, integrated pest management and scientific watershed management can be described in a macro manner. However, they will have little relevance to local farming communities unless they are converted into precise schedules of operations fitting the local crop and farming systems calendar. One of the difficulties in promoting sustainable agriculture is the absence of such location specific data and information. This can come about only through participatory research carried out jointly with farm men and women. Chambers (1988) and Conway and Barbier (1990) have described how this can be done. Chambers (1991) has stressed the need for tailoring technologies to specific growing conditions in complex, diverse and risk prone (CDR) areas.

The promotion of sustainable agriculture will also need specific research tools including genetic resources which provide the genes essential for imparting sustainability to the farming system. We need a special programme for the collection, conservation, evaluation and enhancement of crop genetic resources directly related to the promotion of sustainable advances in crop productivity. Such a specialised Genetic Garden for Sustainable Agriculture would help to assemble material such as the following:

(a) Germplasm of nitrogen fixing tree species and shrubs, stemnodulating legumes like *Sesbania rostrata*, as well as *Azolla* and blue green algae;

(b) Plant species of importance in pest control including tree species, annual plants, fungi and bacteria which act as repellents of pests and those which control soil nematodes and weeds;

(c) Germplasm of species which can enhance the efficiency of fertilizer use including Neem (*Azadirachta indica*) and other tree species whose seed cakes have a nitrification inhibition capacity;

(d) Species which can help to prevent/reduce soil erosion and protect local food security, often referred to as ecological or economic key species, such as vettiver grass (*vetiviera zi zanaides*), *Chenopodium* species, and grain and leaf *Amaranths;*

(e) Tree species and shrubs of value in agroforestry and alley farming practices and in restoring the fertility of degraded and wastelands; and

(f) Plant species of value in veterinary medicine.

Such a specialised genetic garden is currently being established at Madras.

We also need urgently an ocean capability classification and an integrated planning of sea and land surface along the sea coast. Nearly 60 per cent of the human population live within 60 km of coastal waters and two-thirds of the world's cities with population of 2.5 million or more are near tidal estuaries. Within the next 20-30 years the population of the coastal zone is projected to almost double. Over-exploitation has reduced the yield of many fisheries. Competition is also growing among small-scale, large-scale and recreational fishing. The time has come for greater attention to the use of an ecosystem approach for management of marine resources. Coastal aqua culture programmes should not be at the expense of mangrove ecosystems, which include sea grass beds and coral reefs.

Monitoring Sustainability

No single indicator is likely to incorporate the many normative, and therefore difficult to quantify, judgements that must be made in assessing progress towards sustainable agriculture. These would include judgement on issues like reversibility of soil degradation, the level of genetic diversity necessary to ensure crop security and the level of soil microbiological activity necessary for the health of the soil. Nevertheless several quantifiable indicators such as soil organic matter, salinity or acidity, groundwater level and quality and biomass yields per hectare could be used for long term measurements.

If the goal of sustainable development is, as mentioned at the beginning of this chapter, "improving the quality of human life within the carrying capacity of supporting ecosystems," we need tools for estimating the relationship between human societies and the carrying capacity of supporting ecosystems. A suitable tool for measuring this relationship is the Sustainable Livelihood Security Index (SLSI) developed by Swaminathan (1991).

In 1990, UNDP published its first Human Development Report which contained a methodology for the construction of a Human Development Index (HDI). HDI is based on three types of deprivation. These are : life expectancy, adult literacy, and income for a decent living standard. The minimum values were chosen by taking the lowest 1987 national value for each indicator. For life expectancy at birth, the minimum value was 42 years in Afghanistan, Ethiopia and Sierra Leone. For adult literacy, it was 12 per cent in Somalia. For the purchasing power adjusted GDP per capita, the value was US $ 220 in Zaire (Human Development Report, 1990; 1991).

The central message of UNDP's Human Development Report of 1990, to quote Mr. William H. Draper III, Administrator of UNDP, is that "while growth in national production (GDP) is absolutely necessary to meet all essential human objectives, what is important is to study how the growth translates or fails to translate into human development in various societies. Some societies have achieved high levels of human development at modest levels of per capita income. Other societies have failed to translate their comparatively high income levels and rapid economic growth into commensurate levels of human development."

UNDP's 1991 Human Development Report shows that India has a low HDI rank, occupying the 123rd position among the 160 countries studied. The report concluded that the potential is enormous for restructuring national budgets and international aid in favour of human development. Within India, there is obviously considerable variability at the State level in HDI. Kerala, for example, has a high HDI, although the per capita income in the State is lower than in several other States.

Several analysts have suggested a further elaboration of HDI. Lester Brown for example states, "while the HDI represents a distinct improvement over income figures as a measure of changes in human well-being, it says nothing about environmental degradation. As a result, the HDI can rise through gains in literacy, life expectancy or purchasing power that are financed by the depletion of natural support systems, setting the stage for a longer term deterioration in living conditions." Herman Daly and John Cobb (1989) proposed an Index of Sustainable Economic Welfare (ISEW), which takes into account the impact of development on environmental capital stocks. They recalculated the progress of the United States environmental capital stocks, based upon that index and observed that since 1976 there has actually been regress and not progress. Lester Brown prefers per capita grain consumption as an index of human well-being. The amount of money required to purchase the minimum essential calories of food in urban and rural areas has generally been used by the Planning Commission of India to fix the poverty line.

Sustaining and Extending the Green Revolution

The "Green Revolution" technology in wheat, rice, maize and other crops has become associated in the public mind with the use of large doses of mineral fertilizers and pesticides. This is because the semi-dwarf genotypes responded favourably to good soil fertility and good water and crop management, thereby making it economical for farmers to invest in larger quantities of market-purchased inputs. In the early seventies, social scientists feared that the new yield-enchancing technologies may have built-in seeds of social discrimination. This fear has however now led to the realisation that the onward march of the green revolution in wheat, rice and maize and its spread to more areas and farming systems is an ecological and economic imperative in population-rich but land-short countries like Bangladesh, China, India and Indonesia. If this does not happen, the income of small-farm families cannot be increased since they will have very little marketable surplus and thus will not be able to profit from the remunerative output-pricing policies of governments. Also, it will not be possible to prevent further expansion of the arable area at the expense of forests and soils vulnerable to erosion or other forms of deterioration.

Affluent countries have gross national products (GNP) that vary from twelve times the value of agricultural production (e.g. New Zealand) to fifty times (e.g. Belgium, Germany, United Kingdom and the USA). In these countries employment in the tertiary sector exceeds that in the primary and secondary sectors combined. In India, in contrast, over 30 per cent of GNP comes from the farm sector but what is even more significant is that 70 per cent of the current population depends upon the primary sector for its livelihood security. Today, every fourth farmer in the world is an Indian. This situation has led to nearly 47 per cent of the land resources of India being utilised for agriculture. The land available for pastures and grazing is hardly 3.7 per cent, although India also has nearly 20 per cent of the world's farm animal population (cattle, buffalo, sheep, goats, pigs and poultry). Again the contrast in land use with a country like USA is striking (see Figures 10.1 and 10.2).

In global terms, opportunities for traditional rural societies to materially enhance their standard of living are closely linked to the decline in the proportion of the farm population. About half the population of the world today is working or seeking work in agriculture.

Only 10 per cent of the population in the developed countries in 1985 was in the farm sector, shrinking from 38 per cent of the population in 1950. In the developing world, on the other hand, the population working in agriculture was 81 per cent in 1950 and 63 per cent in 1985.

The famine of food at the household level largely arises from a famine of jobs and purchasing power. The pace of progress in the movement of men and women from the primary to the secondary and tertiary sectors has been slow. Self-employment opportunities in non-traditional and off-farm occupations are still limited. On the other hand, population growth has resulted in a growing dimunition and fragmentation of farm holdings.

It is well known that diversification of employment opportunities and income sources insulates people from violent fluctuation in their economic fortunes. In the MINK region of USA (Missourt, Iowa. Nebraska and Kansas), the percentage of farm income to total income is 3.4 times that found in USA as a whole. However, a study by Resources for the Future showed that the economy of region may still not suffer seriously if changes occur in the future in temperature and precipitation as a result of the global warming phenomenon. Since the share of the farm sector in the income of this region is only 3.87 per cent. In India, even now, drought causes serious disruption in the economy of the people and makes it essential for Government to initiate large scale Food for Work programmes. Such a divergent pattern of development has also led to a gradual widening of the gap in per capita income between industrialised and developing nations. For example, in 1880 the per capita ratio of real income between Europe and India was 2 : 1. By 1965, it was 40 : 1, Today, it is over 70 : 1.

Pathways of "Greening" the Green Revolution in Industrialised Nations

There are various prescriptions to achieving ecological farming. Masanobu Fukuoka (1978) recommends the following four steps to make farming natural.

First, there should be no cultivation, that is, no plough or turning of the soil. The earth cultivates itself naturally by means of the penetration of plant roots and the activities of microorganisms, small animals and earthworms.

Second, no chemical fertilizer or prepared compost should be used.

If left to itself, the soil maintains its fertility naturally, in accordance with the orderly cycle of plant and animal life, as is seen in forests.

Third, no weeding by tillage or by herbicides should be resorted to. Straw mulch, a ground cover of white clover interplanted with the crops, and temporary flooding provide effective weed control.

Four, there should be no dependence on chemicals. Nature, left alone, is in perfect balance. For fertilizer, cultivation of white clover, incorporation of straw and addition of poultry manure are adequate.

Moves to reduce the use of chemicals are also a foot in Europe. Scandinavian countries have enacted legislation to reduce pesticide use, with Denmark calling for a reduction of 50 per cent by 1997 and Sweden setting a target in 1988 of a 50 per cent cutback within only 5 years. Their lead has already been followed by other countries in Europe, with the Netherlands proposing a 50 per cent reduction by the year 2000 (Warrell, 1990). The European Consumers Pesticides Charter published in 1990 calls for 20 per cent of European food to be produced using organic methods (Warrell, 1990).

The US National Research Council in a report on Alternative Agriculture (1989) defined alternative agriculture as any system of food or fiber production that systematically pursues the following goals:

(a) Incorporates natural processes such as nutrient cycles, nitrogen fixation, and pest-predator relationships into the agricultural production process;
(b) Reduces the use of off-farm inputs with the greatest potential to harm the environment or the health of farmers and consumers;
(c) Makes greater use of the biological and genetic potential of plant and animal species;
(d) Improves the match between cropping patterns and the productive potential and physical limitations of agricultural lands to *ensure long term sustainability of current production levels* (italics mine); and
(e) Emphasizes improved farm management and conservation of soil, water, energy and biological resources.

The above principles are being packaged into a Low Input Sustainable Agriculture Programme (LISA).

The above example of efforts in industrialised nations to promote the long-term sustainability of current yield levels are indicative of the technology options under consideration. In industrialized nations efforts to limit and stabilise farm production are also motivated by a desire on the part of Governments to reduce the growing subsidy burden, which has reached very high levels in USA, Japan and EC Countries.

Achieving Sustainable Advances in the Productivity of Small Scale Farming

The core of the concept of sustainable agriculture lies in equal and concurrent attention to intra-generational and inter-generational equity. Intra-generational equity, which aims at giving a fair deal to the economically and ecologically disadvantaged farm families, involves attention to issues such as adding a dimension of resource neutrality to scale neutrality in technology development and dissemination, thereby helping all farmers irrespective of the size of their holding and innate input-mobilizing and risk-taking capacity to derive economic benefit from new technologies. Both research strategies and public policies will have to be integrated suitably to achieve this goal. A pro-poor bias must be built into the technology generation system.

Inter-generational equity involves attention to the conservation of the ecological foundations essential for sustainable advances in biological productivity. This will call for added efforts in the conservation of land, water, and biological diversity and in the elimination of agriculture's contributions to the accumulation of greenhouse gases in the atmosphere.

The CGIAR Committee on Sustainable Agriculture in its report submitted at the Consultative Group Meeting held at the Hague, the Netherlands during May 21-25, 1990, arrived at the following conclusions.

(a) No single indicator is likely to incorporate many normative judgements that are difficult to quantity, such as the reversibility of degradation, the critical threshold of decline, or the level of diversity necessary to protect the future genetic base of agriculture; and

(b) Several quantifiable indicators taken over time can provide data along crucial dimensions that help to indicate the sustainability of most agricultural production systems. These include, especially, soil organic matter, soil acidity, crop yields or biomass yields per hectare and net value added to production.

The Committee called for support for developing and testing quantitative models to guide future sustainability research and to extend the utility of findings to new situations.

Where do we go from here?

It is generally agreed that we need technologies which can help to increase the productivity and profitability of small farms without forcing trade-offs between current and future production systems. It is also widely held that techniques like integrated nutrient supply involving a blend of bio-fertilizer, organic and green manures and mineral fertilizers, and Integrated Pest Management (IPM) involving genetic, biological and cultural control methods as well as need-based application of chemical pesticides, will help to combine economic efficiency and ecological sustainability.

Although IPM methodology has been available in crops like rice and cotton for over two decades, its large scale applications under field conditions have been rather limited. This is partly due to the inadequate attention given to the social mobilisation needed for the effective adoption of such technologies. In Indonesia, where group farming practices exist, a major public policy decision to promote IPM led to dramatic results in cutting down the use of chemical pesticides from 1987 onwards. (Fig. 10.3)

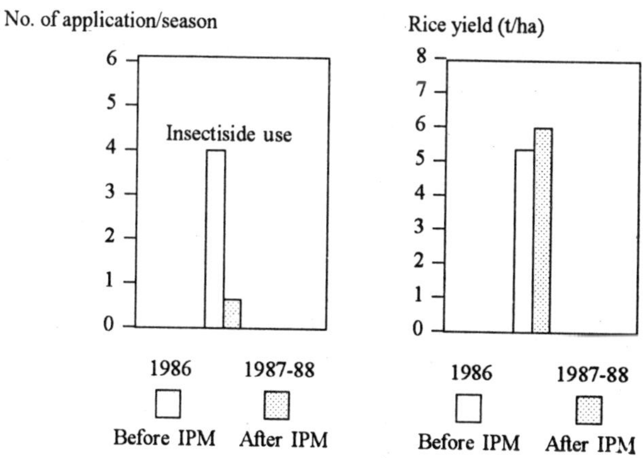

Fig. 10.3 : Impact of IPM on Insectisides Use and Rice Yields in Indonesia, 1986-1988

Source : FAO programme in Indonesia, P. Kenmore, unpublished data.

Any research agenda for sustainable agriculture should give priority to a consideration of the livelihood security of families with either no assets or with small holdings. Chambers (1991) has often stressed that in the case of the green revolution technologies in wheat and rice, farmers who took advantage of them adapted the growing conditions to suit the needs of the technology. On the other hand, farms located in CDR (complex, diverse and risk-prone) areas will need technologies tailored to their agro-ecological and socio-economic circumstances. How can this be done?

Recent experience reveals that varieties with good adaptation to diverse growing conditions can be developed. The "Verry " lines in wheat provide one example. Recombinant DNA techniques are providing additional tools for achieving novel genetic combinations. It is theoretically possible how to breed high-yielding varieties of cereals and other crops possessing multiple resistance to a wide range of biotic and abiotic stresses. Transformation of cells and regeneration of fertile transgenic plants has also been achieved recently in maize (Gordon-Kamm *et al.* 1990). Given appropriate breeding methods and selection procedures, it should be possible to develop strains for CDR areas which, when integrated with efficient methods of soil fertility restoration and water saving and sharing will help to elevate and stabilize yields in such areas.

In many CDR areas, the systems of land use may have to be altered to make them ecologically sound. Agro-forestry involving sylvi-pastoral or sylvi-horticultural practices may be the ideal land and water management system. However, unless the people of such areas are assured of a steady supply of basic staples at reasonable prices, they will continue to grow annual crops in marginal and erodable soils.

Many steps in conservation are perceived by local populations as measures which will erode rather than strengthen their livelihood security. For example, the list of "threatened protected areas" is growing (IUCN, 1990). In many parts of India, protecting the protected areas is a priority task. Landuse changes in ecologically desirable directions can be triggered only through public policy measures designed to ensure economic and physical access to food to the families cultivating the land.

To conclude, the productivity pathway of production improvement is the only option available to countries like India and China for meeting their needs for agriculture commodities. An appropriate integration of frontier and traditional technologies can help to combine the ecological and employment intensity strengths of traditional technologies with the economic efficiency characteristic of modern technologies. Genetic strains which are less dependent on market-purchased chemical inputs are also becoming available. Green manure crops which can fix nitrogen in the stem, in addition to fixation in the rhizosphere, are attracting attention.

Farmers are the ultimate judges of the value of new technologies. The major concerns of farmers with small holdings are first, household food security and second, remunerative marketing opportunity. It will not be possible to achieve the goal of "food for all" within the confines of the carrying capacity of the supporting ecosystems if "green" technologies capable of strengthening the livelihood security of rural families do not become available. In this context, it would be useful to refer briefly to what is being done in technology integration in rice, a major food crop of the world.

Note : Reproduced from *Indian Journal of Quantitative Economics*, pp. 109-129 with kind permission of the publisher.

REFERENCES

Chambers, R., (1988), *Managing Canal Irrigation*, New Delhi : Oxford & IBH Publishing Co.

Chambers, R., (1991), "Reaching the Unreached—Problem of Paradigms?" in Biotechnology : Reaching the Unreached (ed.) M.S. Swaminathan and Vineeta Hoon. Proceedings No. 3. M.S. Swaminathan Research Foundation. Madras, 143-147.

Conway, G.R. and E.B. Barbier, (1990), *After the Green Revolution : Sustainable Agriculture for Development,* London : Earthscan Publications Ltd.

Daly, H. and J. Cobb., (1989), *For the Common God Boston* : Beacon Press.

11

FARM SECTOR REORGANIZATION FOR SUSTAINABLE GROWTH AND OPTIMAL PRODUCTIVITY

A View Point

S.S. JOHL

Although in physical sense the cultivable land area in India is limited which cannot be expanded horizontally. Irrigated or assured water supply area is further limited in acreage. Still, the country has vast span of land resource that should be able to produce enough food for its population and addition to this population at present level of growth for atleast next quarter century, provided production patterns are appropriately adjusted and cropped area is expanded vertically through continuous improvements in productivity. Yet this statement needs to be qualified with the factual position that if the present neck and neck race between food production and human reproduction continues unhindered and untackled, population growth is bound to take over, which will fuel hunger, generate upheavals and will ultimately lead to disintegration of the society beyond control. Thus, any plan or approach that is aimed at intensifying agricultural production must not ignore the fact that checking the population growth is a pre-requisite to the improvement of quality of life in India.

Unfortunately, India's approach to this daunting challenge, if not disquietingly lacking, has remained and still is mute and indifferent. Estimates are that within next twenty years, India's population is going to exceed that of China. The decision-makers and planners in the country need to realize that no amount of research and its application, production intensification and diversification can solve the food problem in this country in the long run on a sustained basis, if the present level of indifference to population planning continues. It is unfortunate that political environment in this country and politician's concern and priorities are not all conducive to the tackling of this problem. In my opinion this agricultural science congress must express its awakened and serious concern at the untackled and unchecked population growth in this country, which is worse than the developing of lethal black hole in the ozone layer, which is a matter of life and death for the whole world. After all it is the population and its growth when added for all countries world over becomes the cause of this disastrous phenomenon.

A more serious development in this respect is that whatever investment in population control and family planning this country has made, it has resulted in deterioration of the quality of the population. This deterioration is now progressing at an exponential rate. Today the situation and trend is that those who can afford to bring up properly and educate adequately their children to be responsible and productive citizens and have the resources to settle them suitably in life, are limiting their families to one or two children. But those who cannot afford one square meal a day and are living near or below the line of poverty, are having no concern for family planning. As a consequence the ratio of the have nots, poor and emaciated anemic population is increasing in the total population. This ignorant unconcerned population is vitiating the total political, social and cultural development of the society and is adversely influencing the policy-making in this country. Every concerned and enlightened citizen, and more so those who are in authority, need to wake up to the reality of situation in this respect. Otherwise this nation sooner than later is doomed to become a basket case, because all the gains in food production and growth in Gross Domestic Product (GDP) in real terms are going to be completely absorbed and negated by the last growing population of the have-not sections of the society.

Given the situation on population front and notwithstanding an urgent, rather dire, need to tackle this problem on top priority, there is an urgency to produce enough food on a sustained basis in order to avert developing shortages and their accompanying negative influences on the very economic, social and political fabric of the society. Emphasis here is on sustained growth in production that is enough to meet not only the future growing demand, but also the need of the population. It is very important to distinguish between Need and Demand. If we calculate the total need for food on the basis of energy requirements of the Indian population, the country is and had remained short of food requirements. But India claimed to be surplus in foodgrains for the past several years and occasionally in earlier years too and even exported wheat as well as rice. India indeed became surplus in supply of foodgrain when considered on the basis of demand for foodgrain. Since demand is always at a price and large proportion of the population, specially the people below poverty line, did not and do not have enough purchasing power to buy the available foodgrain at going price. Hence excess of supply over demand, which renders the concept of foodgrain surpluses a misnomer in this country. On the other side if highly subsidized foodgrain, specially the two rupees per kg., supplied in some states are priced at market rate, even the existing low level of stocks of foodgrains will go surplus. Such stocks, however, cannot qualify to be genuine surpluses or exportable surpluses.

Now coming to the issue of food availability on a sustained basis for future, if the consumers are receptive to some changes in their food habits, the problem of shortages in food availability can be resolved or can be atleast postponed for a long time to come. Humans are bound by received habits that determine their staple foods in different areas and regions of the world. In spite of the fact that given area can produce several times higher levels of food in terms of consumable energy by way of different products or combination of products, populations keep sticking to the staples of their choice developed over generations. This creates food shortages and sometimes famines in several areas of the world. For instance in Southern Ethopia land is so fertile and virgin and the climate year round is so conducive that if wheat and corn are grown, it would make the country surplus in food. Yet the peoples preference which is virtually an addiction, goes for Teff (*Eragristus avtssubica)* which is actually a grass and the best of the crop does not yield more than a couple of quintals per acre. As a consequence the country suffers from serious food shortages. It is basically the food shortage that bred upheavals and disruptions among regions resulting in disintegration of the country. Such examples can be easily quoted from different parts of the world including African, Latin American countries and erstwhile USSR. If the food basket does not change, India would need foodgrains minimum 206 million metric tonnes (measured in terms of demand) by the end of century. Some estimates project this demand to 240 million tonnes under different assumptions of growth and income distribution.[1]

TABLE 11.1 : Net availability of cereals, pulses and other important items of consumption, in India, 1951 through 1995.

Year	*Population (million)*	*Cereals (Million tonnes)*			*Pulses Net Avail-ability (Million tonnes)*	*Per Capita Net Availability/Annum*					
		Net Prod.	*Net Imports*	*Net Avail-ability*		*Cereals (Kgs.)*	*Pulses (Kgs.)*	*Edible oil & Vanaspati (Kgs.)*	*Sugar (Kgs)*	*Cotton (Mts.)*	*Electricity domestic (KwH)*
1951	363.2	40.1	4.1	44.3	8.0	122	22.15	3.2	5.0	14.4	2.4
1961	442.4	60.9	5.5	64.6	11.1	146	25.2	4.0	4.8	13.8	3.4
1971	551.3	84.5	2.0	84.0	10.3	152.4	18.7	4.5	7.4	13.6	7.0
1981	688.5	100.1	0.5	104.8	9.4	152.3	13.7	5.0	7.3	11.0	13.5
1991	851.7	141.9	–0.6	145.7	12.9	171.0	15.2	6.5	12.7	15.1	38.2
1992	867.8	136.8	–0.8	137.6	10.9	158.6	12.5	6.6	13.0	13.7	41.9
1993	883.9	145.8	2.6	138.0	11.2	156.2	12.7	6.9	13.7	15.6	45.6
1994	899.9	149.6	0.5	142.6	11.6	158.4	13.0	7.0	12.4	15.9	48.7
1995	916.0	154.9	–	155.5	12.4	169.7	13.5	7.5	13.0	15.1	53.0
1996	932.0	150.5	1.6	157.8	11.6	169.4	12.4	8.2	14.1	16.3	–

Source : Agril. Statistics at A Glance, Directorate of Econ. & Stac., Ministry of Agri.. 1987, 1994.

Even as of today, when India has run short of wheat and resort has been made to imports, the country is overflowing with surpluses and holding more than adequate stocks of rice (Although even at these levels of wheat stocks there was no need for panic to import, which demonstrates fickle mindedness and lack of adequate understanding of the undercurrents of supply and demand in the national and international market). Advancing the argument a little further, if progressively at the margin food habits develop or are encouraged to develop to include more and more of potatoes and vegetables in the food basket of the Indian families, especially in the poorer sections of the society, there could be lesser danger of food availability running short of requirements for long time to come. Here comes the problem of taste and habit to consume received staples that developed over generations through centuries in the past. The example of Teff consumption in Ethiopia stated above is an illustration of the taste and staple food rigidity in the solution to the problem of food shortage in the human society. An appropriate sustained shift in food habits and composition of consumer's food basket enhances the possibilities of vertical expansion of crop acreage through increase in productivity of food per unit of cropped area per unit of time. This makes it more easily possible to resort to intensive cultivation, without leaving debilitating effects on soil health and water quality, because these non-grain food crops generate and incorporate more organic matter into the soil per unit of time. The higher organic matter content ensures better utilization of chemical fertilizers, making it possible to apply less amounts of chemical fertilizers to the soil per unit of food equivalents. It is a challenge for agricultural scientists to produce appropriately usable cost-effective technology in this direction so that adequate and low cost food becomes available to the fast growing population in this country on a sustained basis without mining the land's productive stock of nutrients, specially the micro-nutrients, to adversely affect its health.

This approach to the sustainability of food production, however, requires development of extensive food processing facilities that ensures storability and non-perishability of the products developed out of these alternative crops. Problem here is that as soon as we shift production from foodgrains such as wheat or rice or for that matter any foodgrain to any other food crop, we are faced with uncertainty and risk in production as well as prices that affect cost effectiveness and incomes from these crops.

Production levels in terms of productivity and also quality of production and processing of these agricultural commodities from produce to product cost-effectively, falls in the purview of production and processing technology, which is the responsibility of the physical and biological scientists as well as processing engineering. The price uncertainty and risk though primarily originating form quality and quantity of production, depend on market development and distribution system so that the products are available to the consumers in the form, at the place and time the consumer demands and at prices he can afford.

Thus production, processing, storage, distribution and market development have to proceed together in an integrated manner and simultaneously, if the country has to measure upto the challenges of providing food security to the growing population on a sustained basis. This requires a change from a system of subsistence farming to commercial production on considerably large scale. What is important is not only the total production, but more so the marketable surpluses the agriculture sector generates, which move out to the industrial sector as wage good. We have often experienced the phenomenon that is spite of higher production of foodgrains, procurements did not increase and off takes from Public Distribution System (PDS) did not decrease. The prices too did not ease. Notwithstanding the inaccurate estimates of production made by the Ministry of Agriculture, which too often go off the mark, the increase in production when it occurs in areas of the country where producers are net consumers and there is a predominance of subsistence farming, this production gets consumed right there, without even reducing their subsidized foodgrain lifting from the PDS depots. As a consequence urban sector consumers do not get much relief. In fact markets go aflush with arrivals only when production increases in the surplus producing areas, which increases procurement, builds up stocks and eases consumer prices in the open market. For instance production increase in Punjab has a high correlation with procurement with 'r' = 0.52 in respect of Wheat and 0.91 for Rice. Corresponding 'r' between production and procurement for rest of India is 0.05 for Wheat and 0.29 for Rice. Similarly, correlation coefficient between production and build up of stocks is 0.57 for Rice in respect of Punjab and 0.022 for rest of India. Thus, whereas it is essential to develop and promote adoption of improved production technology for socially and economically handicapped areas, rainfed areas and food deficit areas, which are distinct categories of farms and farmers, it is

TABLE 11.2 : Changes in production, procurement and stocks of rice

(in '000 tonnes)

Year	*Punjab*		*Rest of India*		*Punjab*		*Rest of Inida*		*Stocks in India*	
	Prod.	*Δ Prod.*	*Prod.*	*Δ Prod.*	*Procure*	*Δ Procure*	*Procure*	*Δ Procure*	*Ist January*	*Δ Stocks*
1980-81	3233		50397		2531		3078		6630	
1981-82	3750	+517	49500	–897	3110	+579	4224	+1146	6210	–420
1982-83	4156	+406	42964	–6536	3252	+142	3795	–429	5440	–770
1983-84	4536	+380	55564	+12600	3273	+21	4458	+663	4690	–750
1984-85	5052	+516	53288	–2276	4293	+1020	5570	+112	7750	+3060
1985-86	5482	+430	58348	+5060	4217	–76	5659	+89	9880	+2130
1986-87	5949	+469	54611	–3737	4378	+161	4778	–881	9450	–430
1987-88	5442	–507	51418	–3139	3365	–1013	3537	–1241	6480	–2970
1988-89	4932	–519	65567	+14149	2859	–506	4870	+1333	4850	–1630
1989-90	6697	+1774	66873	+1306	5003	+2144	6858	+1988	6960	+2110
1990-91	6681	–16	67609	+736	4821	–182	7855	+997	9630	+2670
1991-92	6755	+74	67925	+316	4249	–572	5958	–1897	9290	–340
1992-93	7000	+245	65860	–2065	4905	+656	8143	+2185	9500	+210
1993-94	7620	+620	71350	+5490	5486	+581	8773	+630	11950	+2450

Correlation Coefficients

(a) Between columns 2 & 6 = 0.909

(b) Between columns 4 & 8 = 0.289^{NS}

(c) Between columns 2 & 10 = 0.57^{NS}

(d) Between columns 4 & 10 = 0.22^{NS}

* Significant at 1% level;

** Significant at 5% level;

NS = Non-significant.

Source : Government of India, Agril. Statistics at a Glance, Directorate of Econ. & Stat., Ministry of Agriculture, 1995.

doubly essential to generate farm surpluses, specially of food and foodgrains that can be procured, stored and moved to deficit areas as and when the need arises.

Thus, surpluses of foodgrains from surplus producing areas hold special importance for the food economy of the country. Any additional production in these areas becomes marketable surplus and large part of these are procured at support prices by the Government, which builds up stocks. Any decrease in arrivals and marketable surpluses from these areas leads to more than proportionate draw down in food stocks which often necessitates imports. Last year is a typical example. It is the decline in procurement of over four million tonnes of wheat from Punjab and Haryana which created panic resulting in import of about two million tonnes of wheat. If this much reduction in production had occurred in net deficit states and production in the surplus producing areas had increased or atleast had not declined, this would have been all marketable surplus which could have been procured. There would have not been any perceptible reduction in stocks and no panic to import wheat in the presence of comfortable stocks. It is, therefore, essential that scientists and policy makers must develop and promote state-of-the-art production technology in order to put production of foodgrains and for that matter any other product on a substantial growth path in the surplus producing areas of the country, without of course any let up in efforts in deficit areas.

Further, it is to be realized that surpluses are generated only on farm sizes that are economically viable beyond the size of subsistence farming. More than half of the farmers in the country are net consumers of foodgrains. They cannot produce their full requirements on their tiny plots of land. Again a farm to be viable must grow in size horizontally and/or vertically. A stagnant farm unit remains an unviable proposition that cannot generate surpluses. According to 1991 census, 59 per cent of the farms in India are less than one hectare in size with an average size of 0.40 hectare. Another 19 per cent of the farms are between one and two hectare size with an average size of 1.44 hectare. Thus at least 78 per cent of the farms are incapable of adoption of improved production technology and cannot generate much of surpluses with existing production technology. Majority of such farms have no independent approach and no independent source of irrigation water. They must plant crops or sow seed as the others do in the areas around. They must also adopt the same production, harvesting and marketing technology and harvest the crop at the time others do. For instance in case of rice crop rain water flows freely from higher level fields to lower level fields owned by different farmers. It is not therefore, economical or even advisable for any one or a few farmers to apply fertilizers to their individual field(s). The decision-making per force has to be a social activity hence no independent initiative is possible. These farms also do not have investment resources, nor can they generate any. Thus capital formation and adoption of improved production technology suffers in the farm sector due to these uneconomic and financially starved farm units.

Answer to this problems is (1) to create responsive land market and (2) to reorganize the farm business in corporate farms in areas where intensive or high value crop cultivation can be pursued.

Land Market

In case of land market, provisions need to made wherein leasing in and leasing out of land on long-term basis becomes risk free for the parties. Also, reasonably priced long term bank credit should be available for buying of land against collateral of the land bought and part or whole of the land and immovable assets of the buyer. This will increase the mobility of the owners of small and tiny plots of land, which otherwise remain either uncultivable or are cultivated with traditional and inefficient production technology. This will improve farm size of the enterprising farmers, who can increase intensity of farming and adopt farm practices which will put the farm unit on a sustained growth path.

Farm Organization

In case of corporate farming, land of the adjoining farmers can be pooled to create farm corporations of 500 to 1000 hectare each in areas where vegetable, flower, fruit, and dairy farming can be pursued with advantage. Farmers should receive shares in lieu of their land pooled and other properties and assets they surrender or contribute. These shares should be negotiable and inheritable instruments. In such farm corporations the shareholders should have a prior right to work as they may be capable of doing and that may be required at the farm. The profit from the corporate farm business may be pro rated according to the number of shares held by the individuals and the work hours or

TABLE 11.3 : Distribution of Operational holdings and average size of holdings, according to farm size groups in India, 1976-77 through 1990-1991.

Category	*No. of operational holdings (in per cent)*			*Area operated (in per cent)*			*Average size of operational holdings (in ha.)*		
	1976-77	*1980-81*	*1990-91*	*1976-77*	*1980-81*	*1990-91*	*1976-77*	*1980-81*	*1990-91*
Marginal (< 1.00 ha.)	54.6	56.6	59.0	10.7	12.0	14.9	0.39	0.39	0.40
Small (1.0-2.0 ha.)	18.1	18.0	19.0	12.8	14.0	17.3	1.42	1.43	1.44
Semi Medium (2.0 - 4.0 ha.)	14.3	14.0	13.2	19.8	21.2	23.2	2.78	2.77	2.76
Medium (4.0 - 10.0 ha.)	10.1	9.0	7.2	30.4	30.0	27.2	6.04	5.98	5.90
Large (> 10.0 ha.)	2.9	2.4	1.6	26.2	22.8	17.4	17.58	17.24	17.33
All holdings	81569	89392	105286	163344	162758	165603	2.00	1.82	1.57
	(100)	(100)	(100.0)	(100)	(100)	(100)			

Source : Agril. Statistics at a Glance, Directorate of Econ. & Stat., Ministry of Agri., 1987, 1994.

Note : Figures for operational holdings and area operated for 'All holdings' are given in thousand numbers and thousand hectares, respectively.

work days put in by the members. Such farm corporations will not recognize the right of ownership to any individual piece of land and will develop the farm on scientific basis in order to improve productivity and incomes on a sustained basis. For such farms, research in farm technology, extension education, farm level R&D, capital formation, bulk marketing, storage and processing would become easily possible and economically viable. For development agencies, it will be easy to provide inputs, credit, education and advisory services. Such corporations can develop better villages or townships with all the modern amenities of life. This approach would put the farm sector on a faster growth path and bring in improvements in living conditions of the rural populations. These farm corporations can be organized for special enterprises or appropriate complementary combination of enterprises based on resource endowments of different areas and market opportunities. Once the system gets initiated through education, persuasion and with mutual agreement and demonstrated effectively, it can take momentum and cover small and marginal farms dominated areas extensively. The management of these farms can be entrusted to democratically elected small board of management and subject matter work based small committees that decide upon production patterns, technology adoption, financing, capital investments as well as market development, sales, processing, storage of produce etc. Even if government may provide specialist services and technical management input free of cost, it will be less costly and more productive than the present system of provision of extension services, which is more of a scattered shot, ill aimed and ill targeted. If India wants to get out of the quagmire of poverty and stagnation, these reforms in farm organization can go a long way in this direction. Otherwise food and agricultural product shortages are bound to develop fast and over take our development efforts sooner than later.

Such or a similar reorganization of the farm sector is a must in order to counter the diluting effects of socialistic dispensations the Indian Government have been administering to this sector for the past five decades, which have not allowed the farm economy to grow faster and on sound lines. Today the situation is that most of the production technology developed at experimental farms is not finding its way to the so called farms i.e. small and marginal farms, because of the financial and physical conditions of these farms are not consistent or identical with those on the experimental farms.

TABLE 11.4 : Changes in production, procurement and stocks of wheat

(In '000 tonnes)

Year	Punjab		Rest of India		Punjab		Rest of Inida		Stocks in India	
	Prod.	*Δ Prod.*	*Prod.*	*Δ Prod.*	*Procure*	*Δ Procure*	*Procure*	*Δ Procure*	*Ist January*	*Δ Stocks*
1980-81	7677		28637		3768		2827		7730	
1981-82	8544	+867	28906	+269	4826	+1058	2892	+65	10180	+2450
1982-83	9168	+624	33626	+4720	5177	+351	3115	+223	13010	+2830
1983-84	9422	+254	36054	+2428	5012	−165	4288	+1173	17180	+4800
1984-85	10176	+754	34053	-2001	6151	+1139	4196	−92	20740	+2930
1985-86	10988	+812	36062	+2009	6484	+333	4052	−144	18900	−1840
1986-87	9447	−1541	34873	−1189	4419	−2065	3461	−591	14860	−4040
1987-88	11084	+1637	35086	+213	4749	+330	1832	−1629	7550	−7310
1988-89	11578	+494	42532	+7446	500	+851	3342	+1510	9500	+1950
1989-90	11681	+103	38169	−4363	6740	+1144	4322	+980	13150	+3650
1990-91	12155	+474	42985	+4816	5543	−1201	2209	−2113	11040	−2110
1991-92	12295	+140	43395	+410	4486	−1055	1892	−317	6730	−4310
1992-93	12370	+75	44840	+1445	6495	+2007	6339	+4447	15220	+8490
1993-94	13340	+970	45790	+950	7285	+790	4583	−1756	17780	+2560

Correlation Coefficients

(a) Between columns 2 & 6 = 0.52**

(b) Between columns 4 & 8 = 0.051^{NS}

(c) Between columns 2 & 10 = 0.037^{NS}

(d) Between columns 4 & 10 = 0.045^{NS}

* Significant at 1% level.

** Significant at 5% level,

NS : Non-significant

Source : Government of India, Agril. Statistics at a Glance, Directorate of Econ. & Stat., Ministry of Agriculture 1995.

Extension and Supply Services

So far the major emphasis in respects of extension services, supplies of inputs and financial advances to the farm sector has been on the delivery system. As a consequence, in spite of extensive as well as intensive effort made by the extension agencies, input supply companies and financial institutions, these services are not proving as effective as they should be and consumption of inputs remains correspondingly below optimum and credit disbursement targets of banks are not achieved in spite of intensive efforts made by the financial institutions. The banks are not able to locate viable agricultural projects or proposals that can be financed safely to achieve their targets of priority sector lending in agriculture. These services, supplies and credit are being provided as one pours food over closed mouths. This means we have not endeavored to open these closed mouths with adequate emphasis on developing responsive and demanding recipient system. Agriculture sector land reforms suggested above and development of farm organizations like farm corporations are aimed at making the recipient system responsive to and capable of absorption of the new technology and advisory as well as input supply services.

Market Development

No production efforts can take off or succeed unless appropriate market gets developed to absorb the commodities and products coming out of the farm sector. Many an effort at increasing or changing production patterns have failed at the footsteps of inadequate and undeveloped market and mismatching of demand with supplies coming out of the agriculture sector. The process of development of the farm sector must therefore start from the market end. The market and processing enterprises must develop first, based on matching and preferably contractual production on farm units. The international market with which the national market is getting integrated quite fast under the economic policies of liberalization and globalization, does not recognize and yield and place to the small and uneconomic production units, of the type that are in majority in India. The marketing organisations, the processing enterprises and farming units have to be therefore large enough to be in a position to absorb state-of-the-art technology in order to render the final product cost effective and competitive in quality that is able to withstand and regours of international market.

It is in this context of the system of production, processing and marketing development that the concept of balancing supplies with growing demand in foodgrains and food scarcity will have to be considered. This approach is a bit in variance with popular bandwagon thinking in India, yet is not in conflict with the concept of achieving self sufficiency through domestic production. It is, therefore, essential that physical and biological scientists interact closely with the development economists and policy makers with a view to chalking out an appropriate framework within which research and extension effort in agriculture are matched with the development of business oriented farm organisation and develop agricultural policy that allows the integration of agricultural industry of the country with the international market from a position of advantage.

Note : Reproduced from *Third Agricultural Science Congress,* PAU, Ludhina, March 12-15, 1997 with kind permission of the publisher.

REFERENCE

1. Hanumantha Rao, C.H. and Ashok Gulati, (1995), *Indian Agriculture: Emerging Prospectives and Policy Issues, Strategic issues in Future Growth of Fertilizer Use in India,* ICAR-IFPRI, New Delhi.

INDEX